PRAIRIE DUCKS

Reproduced through the courtesy of Peter Curry,
Winnipeg, Manitoba, from a painting by Peter Scott.

PRAIRIE DUCKS

A Study of Their Behavior, Ecology and Management

By
LYLE K. SOWLS

A Wildlife Management Institute Publication

University of Nebraska Press
Lincoln and London

Publishers on the Plains

Frontispiece by Peter Scott

Black and White Drawings by H. Albert Hochbaum

First Bison Book printing: 1978
Most recent printing indicated by first digit below:
1   2   3   4   5   6   7   8   9   10

Library of Congress Cataloging in Publication Data

Sowls, Lyle K
  Prairie ducks.

"A Wildlife Management Institute publication."
  Bibliography: p. 179
  1. Ducks. 2. Birds—Behavior. 3. Birds—Manitoba—Delta Marsh. 4. Waterfowl management—Manitoba—Delta Marsh. I. Wildlife Management Institute. II. Title.
[QL696.A52S64   1978]   598.4'1   77–14153
ISBN 0–8032–0978–9
ISBN 0–8032–5895–X pbk.

Bison Book edition published by arrangement with The Wildlife Management Institute.

PREFACE

THE YEAR 1946 brought important endings and beginnings to the North American waterfowl program. Here ended an era when duck populations were estimated and discussed in terms of a hundred million or more; here, with sudden reports of unexpected declines, closed a period of ever-expanding waterfowl inventories.

This same year saw the beginnings of broader programs of fact-finding; and the earliest of these researches revealed that there was an urgent need for more fundamental knowledge about ducks. The studies which began in 1946 followed two trails. One segment of manpower and equipment was directed toward a survey of waterfowl populations on the breeding grounds; and the international survey teams established in 1946 examine the status of nesting waterfowl each year from Missouri to the islands of the Arctic Ocean. These cooperative appraisals of breeding waterfowl were concerned with numbers and trends, but the interpretation of the findings hinged upon more knowledge of the life habits of ducks. Thus the studies of Lyle K. Sowls began in 1946 on a nesting meadow at the edge of Manitoba's Delta Marsh. He set upon an exploration of the relationship between the breeding hen and her nesting environment. By 1950, when his study was completed, it was clear that the investigation had yielded a great mass of new information about the homing and nesting behavior of surface-feeding waterfowl.

The survey and life-history branches of the waterfowl program have worked so closely together that the gist of Sowls' original observations, chronicled herein, hardly is new to most biologists in the field. These facts about homing and renesting were applied to waterfowl management policy long before their final publication. That a study starting broadly, without exacting requirements to attain specific objectives, could produce such a volume of useful information also is important to our wildlife program. Budget bureaus and sportsmen's organizations long have shared an attitude of wary suspicion toward "scientific research" in wildlife. Yet research is simply the seeking of new information by examination of source material; and Sowls did nothing more than approach the ducks themselves to find the answers to certain secrets about their nesting behavior. Now any gunner can understand why the breeding hen must be protected on her home range, spring, summer and fall. Since Sowls began these studies, there has been wide expansion of biological research in waterfowl, the most encouraging feature of which has been the participation of many states and provinces which, until 1946, felt that the federal governments of the United States and Canada were obliged to shoulder all responsibilities in waterfowl problems. We must acknowledge the foresight of the Directors of the Wildlife Management Institute and the Trustees of the North American Wildlife Foundation for their willingness to underwrite this kind of an investigation.

Just as this work has presented information useful to management, so, too, it shows the way toward new avenues of study. What is the relationship of the drake to the nesting environment; what is the pattern of renesting in Redhead and Canvasback; where do ducks go when adverse conditions force them to abandon their home range? With wildlife research, as in other fields of investigation, it is just as important to know where to go as it is to examine the ground already covered. The security of North American waterfowl as game birds hinges not only on the willingness of administrators to apply information such as this to waterfowl policy, but upon the ability of students to seek further investigations. There is ample evidence that administrators and students alike already have made considerable useful reference to these Delta studies now presented in formal publication.

H. ALBERT HOCHBAUM

TABLE OF CONTENTS

PREFACE

CHAPTER I THE STUDY 1
The study area 2
Study techniques, finding nests 3
Trapping hens on their nests 5
Plumage marking 6
Colored bands 7
Study data 8
Chapter summary 8

CHAPTER II MIGRATION AND SPRING ARRIVAL 11
Arrival of spring migrants 11
First to arrive 13
Migration and the weather 17
Reversed migrations 18
Abortive migration attempts 19
Resting and feeding places on arrival 20
Activity of transients 21
Sexual behavior 21
Distance traveled and time involved .. 22
Species association during migration .. 23
Flock size during migration 23
Sex ratio of spring flight 23
Chapter summary 24

CHAPTER III MIGRATIONAL HOMING 25
Homing of adult hens 26
Adult hen pintails 28
Adult hen shovellers 29
Adult hen gadwalls 30
Adult hen blue-winged teal 31
Adult hen mallards 32
Summary of adult hen data 32
Locations of second-year nests 33
Adult hens returning several years ... 33
Return of juveniles 34
Nesting of returning juvenile hens 39
Return of juvenile drakes 40
Homing and waterfowl management . 41
Chapter summary 44

CHAPTER IV　HOME RANGE AND TERRI-
　　　　　　　　TORIALITY ....　47
　　　　　　　An example of home range ....　48
　　　　　　　Duration of home range attachment ..　49
　　　　　　　Territory ....　50
　　　　　　　Territoriality in shoveller #47-604004　53
　　　　　　　Daily variation in ditch population ..　54
　　　　　　　Duration of aggressive behavior .....　57
　　　　　　　Absence of aggressive behavior .....　59
　　　　　　　Variations in territorial behavior .....　60
　　　　　　　Reactions of drakes to "dummies" ..　60
　　　　　　　Chapter summary ....　62

CHAPTER V　NESTING TERRAIN ....　65
　　　　　　　Nesting cover types ....　65
　　　　　　　Grain stubble and fallow ....　66
　　　　　　　Grazed pastures ....　67
　　　　　　　Ungrazed meadows ....　67
　　　　　　　Roadsides ....　67
　　　　　　　Phragmites "jungles" ....　68
　　　　　　　Brush and trees ....　68
　　　　　　　Nesting cover and water levels ......　68
　　　　　　　Influence of grazing ....　69
　　　　　　　Influence of wild mammals ....　71
　　　　　　　Water area types ....　71
　　　　　　　Distances from nests to water ....　73
　　　　　　　Land-water pattern ....　73
　　　　　　　Difference in quality of areas ....　75
　　　　　　　Control of undesirable vegetation ...　76
　　　　　　　Agriculture and waterfowl ....　77
　　　　　　　Chapter summary ....　78

CHAPTER VI　NESTING SEASON ....　81
　　　　　　　Description of data ....　83
　　　　　　　Comparison of nesting dates ....　83
　　　　　　　Comparison of nesting seasons ......　84
　　　　　　　Temperatures and the nesting season .　85
　　　　　　　End of the nesting season ....　89
　　　　　　　Chapter summary ....　90

CHAPTER VII NESTING BEHAVIOR 91
Behavior of the laying hen 91
Building the nest 93
Addition of nesting material 94
Building the canopy 94
Rate of laying 95
Time of day for egg-laying 95
Behavior of drake, accompanying hen 95
Dissolution of pair status 96
Behavior of incubating hen 96
Nest desertion 98
Courtship 98
Retrieving of displaced eggs 101
Rest and feeding periods 102
Egg recognition 103
Eggshell carrying 103
Induced eggshell carrying 106
Response to nest moving 108
Disappearance of eggs 109
Chapter summary 109

CHAPTER VIII BREEDING SEASON MORTALITY .. 113
Nest mortality 114
Human disturbance and predation .. 115
Flooding as a cause of nest loss 115
Destruction of nests by flooding 116
Adult mortality 116
Food habits of marsh predators 117
Foods of adult crows 118
Foods of nestling crows 118
Foods of marsh hawks 119
Foods of mink 119
Franklin ground squirrel foods 120
Changes in predator densities 120
Movements of predators 123
Predator control 125
Chapter summary 127

CHAPTER IX RENESTING 129
Previous work 129
Clutch size, first nests vs. renests 130
Appearance, first nests vs. renests 132

Renesting interval 132
Continuous laying 134
Renesting after loss of brood 136
Location of renests 137
Number of unsuccessful hens 137
Renesting and inventory counts 138
Persistence in renesting 139
Chapter summary 141

CHAPTER X   HEN AND BROOD BEHAVIOR 143
Hatching 143
Brood movements 144
Brood reactions to calls of hen 144
Tolling of intruder by hen 147
Hiding of young by hen 147
Hiding of hen with brood 148
Feigning behavior of hen 148
Defense of young by mother 149
Brood habitat 149
Chapter summary 150

CHAPTER XI   AUTUMN BEHAVIOR AND THE
SHOOTING SEASON 151
Signs of autumn behavior 151
Gathering of "thwarted" pairs 152
Duration of the molt 153
Location of molting areas 154
Population build-up in late summer .. 154
Feeding flights to grain fields 155
Composition of hunters' bag 159
Sex and age ratios 162
Crippling loss from hunting 165
Pattern of autumn departure 166
Chapter summary 168

APPENDIX I   ACKNOWLEDGMENTS 171

APPENDIX II   COMMON AND SCIENTIFIC NAMES
OF PLANTS 174

APPENDIX III COMMON AND SCIENTIFIC NAMES
OF ANIMALS 175

LITERATURE CITED 179

INDEX 185

LIST OF TABLES

Table	Title	
1	Summary of study data	8
2	Spring arrival dates of ducks, 1939-1950	12
3	Dates of arrival of flights, 1946-1950	13
4	Number of adult hens known to return	27
5	Approximate survival rate in pintails	30
6	Delta kill and calculated survival rates	33
7	Distances between nests of different years	34
8	Captive-reared juveniles known to return	36
9	Comparison of return of hens	37
10	Chronology of events of shoveller #47-604004	51
11	Cover plant species for 683 duck nests	66
12	Reactions of plants to water conditions	70
13	Four-year nesting record of #47-604004	85
14	Length of time birds sit on dead eggs	97
15	Occurrence of items in crow stomachs	119
16	Items found in nestling crow stomachs	120
17	Frequency of items in marsh hawk pellets	120
18	Foods found at a marsh hawk nest, 1939	121
19	Items found in mink scats	122
20	Clutch size of first nests and renests	131
21	Clutch size in early and late nesting	132
22	Distances between nests of same hen	137
23	Sample counts of sex-ratio of pintails	153
24	Percentages of species in hunters' bags	162
25	Sex-ratios of five species of ducks	164
26	Sex and age ratios of 10,607 ducks	164
27	Sex and age of 1,039 blue-winged teal	165

LIST OF FIGURES

Figure   Subject

Frontispiece

1   The Delta study area .... Facing 2

2   Pintail hen under a nest trap .... Facing 3

3   Marking duck feathers with paint .... Facing 3

4   Relation of arrival points to nest sites 14

5   Relation of temperature to spring arrival 18

6   Shoveller hen #47-604004 .... Facing 34

7   Nest locations of returning hens .... Facing 35

8   Relation of release point to nest sites 38

9   Home range of shoveller hen 49

10 Changes in roadside ditch populations 55

11 Shared areas of gadwall hens 58

12 Flax stubble on an old field .... Facing 66

13 Pintail nest in a depression .... Facing 66

14 Phragmites "jungle" .... Facing 66

15 Pintail nest on a bare pasture .... Facing 66

16 Unpalatable cordgrass cover .... Facing 67

17 Heavily grazed slough edge .... Facing 67

18 Ungrazed slough edge .... Facing 67

19 Distances from water of duck nests 74

20 Comparative populations on two areas 75

21 Nesting chronology for three species 86

22 Nesting chronology for two species 87

23 Relation of temperature to egg laying 88

24 Shoveller reaches for nest material .... Facing 98

25 Shoveller tucking material into nest .... Facing 98

26 Shoveller reaching for live grass .... Facing 98

27 Shoveller pulling down live grass .... Facing 98

28 Shoveller beginning to retrieve egg .... Facing 99

29 Shoveller completing retrieve of egg .... Facing 99

30 Shoveller flying with eggshell in bill .... Facing 99

31 Pintail hen with eggshell in bill .... Facing 99

32 Duck nest destroyed by a crow .... Facing 114

33 Upper layer of a pintail nest .... Facing 115

34 Lower layer of a pintail nest .... Facing 115

35 First nest of a pintail hen .... Facing 130
36 Second nest of a pintail hen .... Facing 130
37 Third nest of a pintail hen .... Facing 131
38 Regression of renesting interval 134
39 Reproductive tract of laying mallard .... Facing 138
40 Locations of first nests and renests .... Facing 139
41 Brood of blue-winged teal ducklings .... Facing 144
42 Same teal eight hours after hatching .... Facing 145
43 Waterfowl population curves for Delta 156
44 Species composition of Delta bag 160
45 Lesser scaup in hunters' bag 163
46 Labrador retriever with ducks .... Facing 163

CHAPTER I

THE STUDY

*F*ROM April until November there are ducks on the marshes of Southern Manitoba. Along the marsh borders and open meadows, in the small glacial-formed "potholes," on the man-made roadside ditches, and wherever dry land and water meet, one can find the surface-feeding ducks: mallards, pintails, gadwalls, shovellers and blue-winged teal. The diving ducks, the canvasbacks, redheads, ruddies and lesser scaup require a different habitat, however, and can be found deep in the marsh and in the cattail- and bulrush-rimmed sloughs

This book is a report primarily on the five species of surface-feeders. Most of the data contained in these pages were gathered in the Delta marsh between 1946 and 1950. A small portion of the nesting-cover data and the nest-loss data was gathered there in 1939 and 1940.

The study was an intensive one conducted principally through the observation of marked individual birds. A total of 2,181 ducks were banded. Some of these wore colored bands in addition to numbered U. S. Fish and Wildlife Service aluminum bands so that they could be identified as individuals when they stood on mud-banks or floating logs. Still others were marked with paints so that they could be identified easily in flight. Although some of these banded birds never were seen or heard of again after they were released, the fate of others was known from band returns. Many were shot near the place where they were banded while others were taken in hunters' bags in Illinois, Minnesota, the Dakotas, Texas, far-off Cuba, Columbia, Panama and many other places. Some of the banded birds were seen almost daily for many weeks following banding; and still others returned to the same marsh each spring for several years and became long-term contributors to this project.

The broad purpose of this study was to gather basic information on waterfowl behavior and ecology which would be of value to waterfowl management. The objectives of the study were: (1) to obtain basic information on renesting of waterfowl and to study renesting behavior, (2) to examine the strength of migrational homing in a breeding population, (3) to study the land-water pattern and the cover requirements of the surface-feeding species, (4) to measure the influence of spring weather on the breeding cycle, and (5) to study the movements of a resident waterfowl population on its breeding ground.

The study area.—The Delta marsh lies at the south end of Lake Manitoba, one of the three large lakes now remaining on the floor of Glacial Lake Agassiz. The lake is long and narrow, being about 115 miles in length and varying at its southern end from a few miles to 30 miles in width. It is shallow and has a sandy bottom. Through the years, ice and wind have heaved against the southern shore, building a long sandy beach, and trapping part of the lake behind it. The trapped part of the lake and the surrounding marsh, an area of about 30,000 acres, is known today as the Delta marsh. This great marsh is an expanse of meadows, sloughs, bays and channels. In some places it is a few hundred yards wide; in others its width extends for several miles. A sandy, tree-covered ridge forms its northern border.

The large bays and sloughs in the Delta marsh are bordered by hard-stem and soft-stem bulrush, cattail and phragmites.[1] In their shallow waters grows a variety of pondweeds. These areas form the nesting grounds for the diving ducks. The drier meadows are favored by the surface-feeding species.

On the south, the marsh is bordered by farmland. The marsh's border on the south, however, is not so sharp as its northern edge. Here farmland juts into the marsh and the marsh into farmland. In dry years farmers invade the marsh to cut hay, sow a field of flax, or run "dry stock." But in wet years the farmers retreat. High alkalinity is evidenced by a

---

[1] See Appendices for scientific names of animals and plants mentioned in the text.

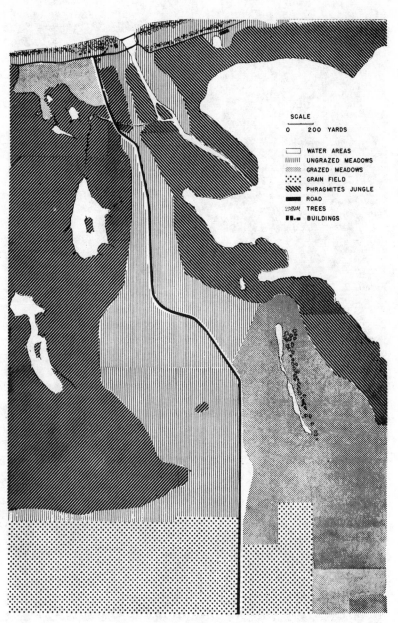

SCALE

0   200 YARDS

☐ WATER AREAS
|||||| UNGRAZED MEADOWS
░░░ GRAZED MEADOWS
∴∴∴ GRAIN FIELD
⧵⧵⧵ PHRAGMITES JUNGLE
▬ ROAD
🌲 TREES
▮▮▬ BUILDINGS

Figure 1. The Delta Study Area.

Figure 2. Incubating hens were captured on their nests by use of a drop trap set in place over the nest and tripped from a distance.

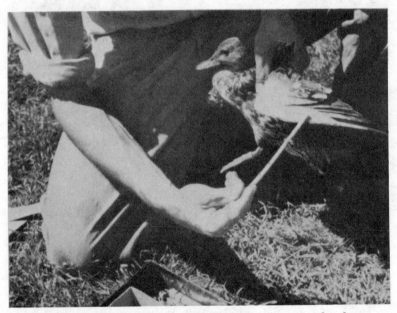

Figure 3. Airplane-dope paint was spread on wings and tail to identify individuals.

white crust on black soil, and glasswort makes a sparse cover over the ground. Farther south the soil is excellent; and wheat, barley, and oats wave high, and produce bumper crops.

To study the ecology and behavior of the surface-feeding species of ducks, I selected an area of about 3,115 acres, transecting the marsh from north to south. This tract, shown in Figure 1, was about 2½ miles long, 2 miles wide, and ran in a north-south direction from the south shore of Lake Manitoba at the village of Delta to the prairie farmlands on the Portage Plains. Through its entire length ran a public gravel highway and a narrow roadside ditch. There was a line of telephone poles along the highway and these poles were numbered to serve as reference points.

About 41 per cent or 1,277 acres consisted of tall yellow cane. Another 19 per cent or 592 acres were water. It included part of a large bay, six potholes, a long narrow roadside ditch and a marsh channel. Ungrazed nesting meadows, grazed pastures and bordering submarginal grain fields made up the remaining 40 per cent or 1,246 acres. It was in this area that most of the study described on the following pages was conducted.

The dominant vegetation on the ungrazed nesting meadows was whitetop, sow thistle, Canada thistle and pigweed. The grazed pasture consisted of bluegrass and cordgrass.

Study techniques, finding nests.—Most nesting studies in the past have used the technique of finding nests by systematic search. When this method is used, a man or line of men walk through nesting cover to flush hens from their nests. It is a sufficiently good method when only a sample of nests is wanted, but for complete coverage it requires the support of other methods because: (1) systematic search fails to reveal nests when the hens are away, (2) some hens "sit tight" and allow a man to walk by them, and (3) nests may be begun and destroyed between searches.

In the Delta study additional methods were employed. These were: (1) the use of a dog to find nests by scent, (2) the dragging of a long rope between two men, (3) the dragging of a long rope between two cars, and (4) the use of a flushing bar along roadside ditch banks. The above methods expedited nest finding, reduced the amount of labor required and increased the percentage of nests found.

The dog used at Delta was a Labrador retriever. Nests usually were discovered as the hens left the nest. The nests of laying hens had enough scent to attract him, however, and in three cases, nest forms without eggs were discovered by the dog.

The technique of rope-dragging between two men was advantageous on still days where vegetation was light. Where cover was heavy, and on windy days when the rope made little more disturbance than the wind itself, the hens sometimes allowed the rope to pass over them.

Rope dragging between two cars, when the ground was solid, was used to good advantage. The efficiency of such dragging was increased when tin cans, in series of twos, were fastened to the rope at 15-foot intervals. The cans made enough noise to flush hens which might have allowed a rope to pass over them. This method could not be used when there were bushes, stumps, rocks or other objects which could hold up or break a rope. I found the technique valuable in reducing the time required to search large areas of grazed pasture where nesting density was low. An observer in the truck-box determined the locations of nests by watching the hens as they flushed.

On one occasion it was important to find the nest of a baldpate that was nesting somewhere in a 200-acre pasture. After all other methods had failed, we dragged the pasture with a rope between two trucks and located the nest in two hours.

At Delta, the roadside ditch bank was a favorite nesting place for blue-winged teal. For this situation, I found a flushing bar to be a valuable time-saver. A 10-foot pole was attached to the rear of a truck by a swivel and was held in place at a 90-degree angle to the midline of the truck by means of ropes. The bar was free to swing down over the sloping bank as the truck moved ahead. To the pole were fastened 6-foot lengths of baling wire at intervals of two feet and to each wire pairs of tin cans were attached. An observer watching from the truck-box then could see ducks leave their nests as they flushed in front of the bar. By this method both sides of a 2-mile roadside ditch could be covered in two hours, a stretch that would require eight to ten man-hours of walking.

In addition to systematic searching and the above methods,

I located nests by watching and interpreting the early-morning activities of laying hens. This consisted of posting a watch at particular vantage points on the study area and noting the locations where hens dropped into the grass following flights from water areas. These hens invariably were accompanied by drakes as they returned to their meadows to lay. After they had settled in the grass, their drakes returned to their waiting places. This technique could be used only during the early morning hours. At Delta, I found that most laying hens went to their nests between sun-up and four hours thereafter. An observer with an 8-power binocular could watch an area of approximately a half-mile radius. It was important to have landmarks to aid in recording nest sites. On the study area I numbered the poles of an electric power line as well as the roadside telephone poles. These two lines ran at right angles to each other and gave reliable reference points for mapping the locations of nest-sites.

After the approximate location of a nest had been mapped, the exact place was determined by the usual methods of search. This usually was accomplished by working the dog in an ever-increasing circle from the spot where the hen was seen to drop into the grass.

Hens varied in the time they required to go to their nests. Sometimes they lit near it. More often they dropped 50 yards from it and approached by short flights, walking the last few feet. This delay in settling down made it necessary to postpone nest hunting until an hour after early-morning watching had ceased.

This technique was useful particularly when the nest of a marked individual or an uncommon species was sought.

Trapping hens on their nests.—In order to work with known individuals, it was necessary to mark the birds. This, of course, hinged upon their capture. Although several means for catching nesting hens had been devised, two methods were employed at Delta: (1) nest-trapping and (2) hand-netting. Using these two methods, I caught 220 hens on their nests during the five seasons of the study.

The nest trap has been described and figured in a previous paper (Sowls, 1949:262) and is shown here in Figure 2. It was patterned after a trap developed by the late Robert Harris who used the drop trap in catching hens (Hochbaum, 1944: 158). Our trap consisted of a frame 2 feet square of ⅜-inch iron rod over which was attached ½-inch-mesh chicken wire. Lead weights were secured on the front of the trap to make it drop heavily. One side of the trap was propped with a tripping stick while the other side was held firmly in place by two hinged rods pushed into the earth. To the tripping stick was attached a long twine which was laid out on the grass so that the trap could be pulled from a distance.

The trap was set over the nests of incubating hens at the time the nests were found. After waiting for two hours, a reasonable time for a hen to return and settle down, the watcher pulled the trigger string. Trapping success was high in most species of ducks, but low in mallards, which proved suspicious and wary. The main shortcoming of the trap was its failure to catch laying hens.

A laying hen seldom returned to a nest over which a nest trap had been set. To capture hens in this period of the nesting cycle I used a large hand net with moderate success. It consisted of a 2-foot circle of ⅜-inch iron rod attached to a 5-foot pole and covered with netting. Its success depended upon the stealth of the operator in approaching the nest and precision in placing the net squarely over the hen. The value of this kind of trap also varied with weather conditions and cover types. In damp, windy weather when the noise of an approaching person was muffled, it worked fairly well. Nests located along the roadside ditch banks were approached easily. In open stretches of cover, where dead vegetation crackled loudly underfoot, a quiet approach was impossible and trapping was difficult. When two small sticks were placed 18 inches from a nest on the north and south sides, its exact position was indicated. This aided in placing the net over the hen.

Plumage marking.—Quick-drying paints have been used by many workers in attempts to mark birds for field identification. At Delta, I tried a variety of paints and found "airplane dope" very successful. This is a quick-drying paint used on model airplanes. In marking waterfowl, I found that either the tail feathers or the outer half of the primaries could be painted. Certain precautions in using paint were necessary so that the feathers did not stick together or become too heavy. I applied the paint to the primaries with a dry coarse plant stem. By holding the wing in a stretched position, I could run the plant stem over the length of the primaries and cover the surface. The primaries were held in this position until the paint was dry (Figure 3). This required a few minutes' time. If the color did not show up well, another coat was applied, care being taken not to get an overload of paint near the ends of the primaries. The technique worked better on the larger ducks—mallards, pintails, and gadwalls—than it did on the smaller blue-winged teal and shovellers. Early in the study it was noticed that the smaller species could not carry the heavy weight of paint on their primaries with ease. For this reason, the paint was put on the proximal end of the primaries, and care was exercised to avoid putting it on too heavily.

I found that white, yellow, and red proved satisfactory for marking waterfowl for field identification in renesting, and for behavior studies. When marked properly, the birds in flight could be identified with an 8-power glass at a distance up to 500 yards. The marking lasted for a period of about two months.

When red, yellow, and white paint were used on wings and tail, enough combinations were obtained for each sex and species of waterfowl. Some of these were easier to see on a flying bird than others. In the renesting study, hens usually were identified as they flew off the nest in front of us. Birds whose identifying colors were not seen clearly on flushing often could be observed at a later time when they were flushed again from the same spot.

Colored bands.—Colored celluloid bands were attached to the legs of birds for more permanent marking. The weakness of this method was that the bands sometimes were lost. For example, of 30 pintails released near Delta, five had lost their bands within six weeks. On the other hand, a wild shoveller hen, banded by me in the spring of 1947, still was wearing her colored marker in 1950 after six migration flights.

Despite the shortcoming, the colored bands now available are indispensable aids in the identification of individual birds through one season and often from one season to the next.

The efficiency of colored bands was increased when ideal

"loafing" places were created. On the nesting study area I could observe banded birds only when they were standing on the edges of grazed potholes, sandbars, or clear ditch banks. Where natural situations were lacking, I anchored floating logs on small water areas. In this way I was able to make frequent records of the color-banded birds.

Study data.—Table 1 gives a summary of study data showing the number of nests found and number of birds in various categories which were banded or banded and color-marked. Of the 593 nests found, many had hatched or were destroyed before the hen was captured. This explains the difference between the number of nests found and the number of hens nest-trapped and marked. Also, the number of nests found was greater than the number of hens found because some nests represent renests of a single hen or the nest of the same hen in a different year. Figures include all species.

TABLE 1. SUMMARY OF STUDY DATA

Year	Nests Found	Hens Marked	Drakes Marked	Ducks Reared and Released	Ducks Banded in August and September
1946	25	4	0	0	140
1947	41	10	0	0	0
1948	106	51	0	133	651
1949	231	91	4	352	520
1950	190	64	33	128	0
Total	593	220	37	613	1311

A total of 2,181 ducks were banded in this study. It is upon observations of these banded birds primarily that the data in the following chapters are based.

Summary

1. The objectives of the study were: (1) to obtain basic information on renesting of waterfowl and to study the renesting behavior, (2) to examine the strength of migrational homing in a breeding population and evaluate its significance to management, (3) to study the land-water pattern and cover requirements of the surface-feeding species, (4) to measure the influence of spring

weather on the breeding cycle, and (5) to study the movements of a resident waterfowl population on its breeding grounds.

2. Most of the data to be described were collected on a study area of 3,115 acres in the Delta marsh in southern Manitoba.

3. Nests were found by systematic search, dragging of ropes over nesting meadows, use of a Labrador retriever dog, and by watching the early-morning movements of hens.

4. Ducks captured were banded with U. S. Fish and Wildlife Service aluminum bands and colored celluloid bands. The wings and tails of some ducks were marked with a quick-drying airplane-dope paint.

5. Marked individual ducks were identified on the wing by the use of binoculars. Colored bands were of value when banded individuals stood on bare loafing banks and loafing logs.

6. During the 5-year period of this study, 2,181 ducks were banded. Of this group 220 were hens trapped on the nest, 37 were drakes trapped during the breeding season, 613 were hand-reared juveniles that were banded and released, and 1,311 were ducks that were captured and banded in late summer.

CHAPTER II

MIGRATION AND SPRING ARRIVAL

*A*MONG the seasonal events on the northern prairies none is more prominent in the minds of men than the return of waterfowl. In the prairie provinces of Canada discussions about ducks are as characteristic as talk of crops and weather. They voice the hope of the winter-weary who look to the land, to seeding time, and to an abundant summer.

But the return of waterfowl does not mean the end of winter weather. Lakes and sloughs still hold their frozen edges. Large snowdrifts bury the feet of scrubby timbers in the prairie bluffs. Winter still is championed by occasional snowy owls perched on haystubs and fence posts. And snow bunting flocks still tumble and roll across the fallow soil and temporary water sheets.

Here on the shallow ponds, before spring has laid true claim, the first ducks and geese will rest. Mallards, pintails and honkers, spear-heading the advance, will be followed, as the temperature rises, by baldpates, teal, scaup, shovellers and gadwalls.

Arrival of spring migrants.—During an extremely early spring, the first mallards and pintails arrived at Delta by the last week in March. During a late spring, however, they were seen about three weeks later. Sometime within this three-week span one could expect to see the first mallard and the first pintail within a few days of each other. As the main flight occurred, the two species were seen together.

TABLE 2. SPRING ARRIVAL DATES OF FIVE SPECIES OF
DUCKS AT DELTA, MANITOBA, 1939-1950 (Compiled by
H. A. Hochbaum and L. K. Sowls)

Year	Mallard	Pintail	Gadwall	Shoveller	B-w Teal
1939	Mar. 29	Apr. 3	Apr. 22	Apr. 23	Apr. 26
1940	Apr. 7	17	23	21	22
1941	4	8	12	10	27
1942	Mar. 23	3	25	12	21
1943	Apr. 3	5	17	13	21
1944	4	10	17	15	19
1945	Mar. 20	Mar. 20	24	27	27
1946	20	20	12	12	12
1947	Apr. 2	Apr. 2	26	9	28
1948	10	10	15	15	20
1949	2	7	21	9	20
1950	17	19	May 9	22	May 3
First	Mar. 20	Mar. 20	Apr. 12	Apr. 9	Apr. 12
Average	Apr. 2	Apr. 5	21	15	23
Last	17	19	May 9	27	May 3
Standard Deviation in Days	9.0	9.0	7.0	6.0	5.0

The dates of the first arrival of surface-feeding ducks at
Delta over a 12-year period are given in Table 2.

The later-arriving shovellers and blue-winged teal did not
show so wide a span of variation in arrival dates. The gadwall,
which did not occur in large numbers at Delta, also was a late
arrival, but its arrival date was not easy to determine because
of the comparatively few birds.

The tendency for early migrants to be less consistent in
arrival date than later migrants is well known among other
kinds of birds. In regard to this point Hickey (1943:24-25)
says, "It is well known, for instance, that the first spring migrants vary widely in their arrival dates from year to year.
From 1923 to 1927, I saw my first migrant robins around New
York City on March 11, March 12, March 15, January 30 and
February 26. As spring progresses, birds arrive with more and
more regularity. Going through my notebook for an extreme

example, I notice that on successive years my first redstarts were seen on May 1, May 1, May 2, May 10, and May 6."

Although we recorded the date of first individuals as accurately as possible, it was apparent that the date of the main flight had more significance in the picture of spring arrival. A lone bird frequently was sighted several days before another individual was seen, the main flight occurring a week or more later.

The main flight of other species was comparatively inconspicuous and therefore more difficult to record. First arrivals coincided closely with the main flight of mallards and pintails.

A comparison of first arrival dates and dates of main migration of mallards and pintails for the years 1946 through 1950 is given in Table 3.

TABLE 3. DATES OF FIRST ARRIVAL AND MAIN FLIGHT OF MALLARDS AND PINTAILS FOR THE YEARS 1946-1950 AT DELTA, MANITOBA

| Year | Mallards | | Pintails | |
	First Arrival	Main Flight	First Arrival	Main Flight
1946	Mar. 20	Apr. 10-15	Mar. 20	Apr. 10-15
1947	Apr. 2	16-20	Apr. 2	16-20
1948	10	21-23	10	21-23
1949	2	10-15	7	10-15
1950	17	19-30	19	19-30

First to arrive.—Do the residents arrive on the breeding ground before the main flight of ducks passes through the area? In other words, are the first ducks to be seen each spring at Delta those that will nest in Manitoba, or are they the transients with the farthest to go? In the spring of 1950 I watched closely to see whether there was a mass movement of birds through the region before our resident birds arrived. The first lone pair of mallards was seen circling a nesting meadow within the study area on the evening of April 17. The hen was quacking noisily and exhibiting the behavior I had come to associate with the seeking of a nest site. Although the hen and her drake were not marked, further observation indicated that a hen nested close to the area where the pair was seen that spring evening.

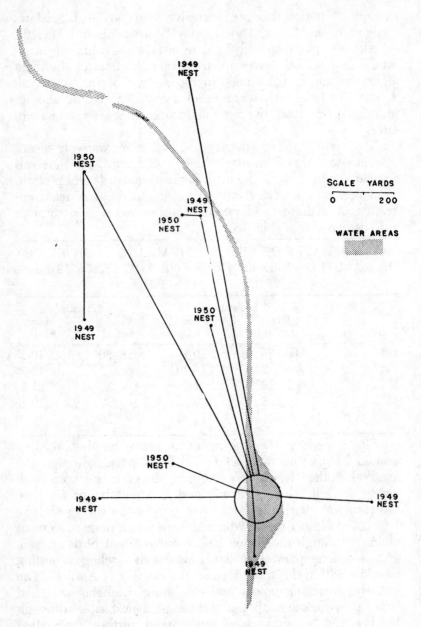

Figure 4. Relationship of 1950 spring arrival point to 1949 and 1950 nest sites of six banded pintail hens.

My observations of marked pintails resulted in further evidence that resident ducks precede the transients to the Delta marshes.

In 1950 the first pintail to be seen in the Delta region, which includes a wide section of the Portage Plains, appeared on a temporary pond in the study area on April 19. At 9:00 a.m. of that day one pair of pintails was recorded. Two days later two pairs of pintails were seen at the same place. On the fourth day there were five pairs. On the fifth day there were 19 pairs and on the sixth day there were 25 pairs. On the night of April 24 the edges of the temporary pond froze over. On April 25 the birds were standing together in the same spot on thin ice, and their hitherto hidden legs were in full view through my 20-power spotting scope. Six banded pintail hens were among them, birds whose history was known for the previous nesting season. They had nested on the study area in 1949, and the next two months were to show that four of them would be found on nests again. The main flight of mallards and pintails did not pass through the Delta region until April 30, and no other large concentrations of birds were seen on the prairies in the meantime. Hence the resident adults were the first to arrive.

The locations of the nests of these birds in relation to their arrival point in 1950 are shown in Figure 4.

In the spring of 1951, following the completion of my study, H. A. Hochbaum observed the return of pintails at Delta. In a letter dated December 7, 1951, Hochbaum says, "Although the first pintails reached the Portage Plains on April 9, 1951 they held to the fields south of the marsh and did not cross to the shoreline of Lake Manitoba nor were they seen along the north edge of the marsh which remained frozen through the first three weeks of April. The first ducks near the station were seen on April 23rd and the first of these on the study area. This first day we saw three pintail and two mallard defense flights; and on the ice in the middle of Rutledge's pasture there was a flock of pintails, four pairs and an extra drake. As we approached, all birds flew away except one pair; the hen had an aluminum band on her left leg and a red plastic band on her right leg. That same evening Norman Godfrey saw two pairs of pintails standing at the edge of the station pond, each female with a red band on its right leg, each male unbanded. These were the first wild birds to come to the pond. I saw the two pairs the next day: the hens were very tame, the drakes were very shy. On first sight, the two pairs were together; but within three days, they usually remained apart on the pond. The two hens remained through the breeding season, the nest of one being located not far

from the study pond. At 6 p.m. of the 24th, the second day of ducks at Delta, there was a pair of pintails on the pond. The female had a yellow band on the left leg, aluminum band on right. At 7 p.m. the same evening another pair was noted with the female wearing an aluminum band and a yellow band on her right leg. . . . No doubt about your marked birds being with the very first arrivals."

Migration and the weather.—At Delta the arrival of the summer resident ducks and the passage of migrants through the region were influenced greatly by weather. Here *we were concerned with the factors affecting a population which already had been on the move for about two and a half months, and not, of course, with those initiating migration.* Here migration fronts were flexible and changing continually with temperature and barometic pressure. The mass of northward moving water-fowl pushed against a barrier that was giving way gradually but inevitably. Sudden changes in the barrier brought sudden spurts of movement of the population; when the barrier of ice and cold air masses increased, the migration sometimes retreated.

Much detailed study of the effects of temperature on bird migration has been made. Nice (1937:45) showed a close cor-relation between the arrival of the first wave of song sparrows at Interpont and the occurrence of warm weather but showed that late-comers were influenced less by temperature. Lincoln (1939), in his early work on migration, showed the movements of various spring migrating birds in relation to spring tempera-tures and drew lines of migration which corresponded to cer-tain temperature lines or isotherms. For example (p. 31), he described the spring flight of the Canada goose and said that the northward migration keeps pace with the advance of the isotherm of 35 degrees Fahrenheit.

The mallards and pintails apparently move northward as fast as open water is available in the manner described by Lincoln, temperature being the dominant external factor af-fecting their migration.

The arrival dates and migration patterns of all species at Delta are not equally well known. It was much easier to study the migration of the abundant mallards and pintails, for ex-ample, than it was the less common baldpate and green-

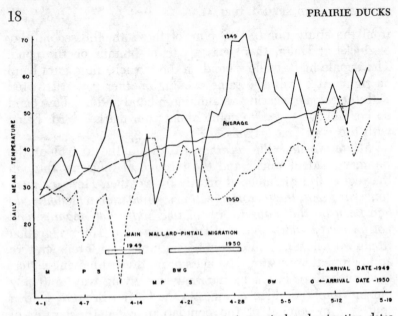

Figure 5. Relationship of temperature to spring arrival and migration dates during the years 1949 and 1950 for mallard, pintail, gadwall, shoveller and blue-winged teal.

winged teal. The effects of temperature on the arrival and main passage of the more common Delta residents and transients are shown graphically in Figure 5. Here a comparison of spring arrival dates of five species in 1949 (an early spring) and 1950 (a late spring) is made. The weather data for this figure are based on information given by the Dominion Government Department of Transport, meteorological division of the Air Services at Winnipeg. Temperature readings were made at the Winnipeg station, 70 miles east of Delta. Arrival dates are for the Delta region including the Portage Plains immediately south of the Delta marsh. Although some difference in temperature may have existed between Winnipeg and Delta, I believe it was too slight to have significance in these comparisons. The average temperature rise was computed for a period of 65 years. In no single year did such a steady rise in daily mean temperature occur. Instead, the newly arrived ducks were exposed to constantly changing temperatures. Where open water existed one day, there was ice the next.

Reversed migrations.—In Figure 5, we see that temperatures fluctuated, in early spring, between thawing and freezing. Thus, loafing flocks of ducks sometimes were forced to retreat to favorable areas. Hochbaum (1944:14) says that ducks forced from the Delta marsh by cold weather retreated to open holes in the Assiniboine River 20 miles to the south. One such mass movement out of the marsh was seen in the spring of 1950. On April 17 the first mallards arrived, and on April 19 the first pintails were seen. From then on there was open water in the marsh; mallards and pintails were scattered throughout the marsh in fairly large numbers. In some sections of the flooded prairies large flocks of transients were loafing and feeding. On the morning of April 26, seven pairs of mallards, 14 pairs of pintails and one pair of baldpates were resting on a large flooded field on the study area. That night the temperature dropped to 25 degrees and left only half of the pond open. The ducks remained there during the day. On the night of April 27, following the heavy freeze, about 3,000 mallards and pintails were seen leaving the marsh and flying south over the prairie. This observation was made by Murry Werbiski, a reliable observer. On April 28 the entire population had disappeared from the study area and very few ducks remained in the marsh. With the return of warm weather and open water on April 30, three days later, the birds were back, and the same group of individual pintails, as determined by their leg bands, were on the study area again.

Abortive migration attempts.—As Hochbaum (1944:16) has pointed out, incoming arrivals move with a south wind. Only few exceptions to this behavior occurred and these were explainable. When held up for long periods by poor weather, transient ducks became restless and made attempts to migrate. The attempt failing, they returned to their resting place.

This behavior was most noticeable among diving ducks but occasionally was seen among puddle species. Lesser scaups that had come to the big bay near the Delta hatchery were seen starting off over the lake-ridge of trees to the north, were gone for a few minutes only, and were seen returning to the bay from which they had just come. On the evening of April 23, 1950, I observed a puddle duck migration which apparently was thwarted by a cold north wind. Four flocks of pintails and mallards ranging from 10 to 40 birds were seen coming over high from the south. After heading out over the lake

in a northerly direction for about three minutes, all flocks turned back toward the marsh, came in, and settled down.

The effects of fog on the migration of birds can be observed easily where local fog banks roll across a body of water. On April 16, 1949, I saw a flock of 30 mallards migrating across the Delta marsh toward the lake. A fog bank was hanging over the shore and extended unbroken over the big lake. The mallard flock entered the fog and thereupon suddenly was dispersed. After flaring and wheeling back, the flock reformed, returned to the marsh, and settled down.

Regardless of wind and temperature, I have never seen spring flocks migrating over the Delta marsh when a fog prevailed.

Resting and feeding places on arrival.—Throughout the prairie provinces of Canada, the first open water occurs in the fields. These ponds, made by melting snows, are temporary; most of them disappeared by June. Here broad, shallow sheets of water cover old grain stubble; and waste grain, along with weed seeds, offer food for transient waterfowl. The bare edges make ideal loafing areas where ducks can rest.

When the previous summer has been dry, the temporary ponds soon disappear into the earth, and the ducks are forced to use small openings in the great marsh. But when the previous summer and fall have been wet, the ground is saturated, and prairie ponds persist longer.

During the springs of 1946, 1947, 1948 and 1949 nearly all of the large transient bands in the Delta area were found on the flooded parts of the Portage Plains. In the spring of 1950, however, large flocks of transients crowded into the first open water on the Delta marsh. This cannot be attributed entirely to an absence of flooded areas on the plains. The particular areas which held large numbers of transients in 1949 still were available in 1950 and appeared to be in the same condition as they had been in 1949. They did not, however, attract ducks in 1950; and I have no information to explain this apparent inconsistency.

Activity of transients.—Activity of the transients during the spring passage was confined largely to feeding, loafing and courting. During this period, mallards and pintails sometimes made daily or twice-daily trips to old grain stubble to feed. In the Delta region I did not see other species make spring stubble flights.

The loafing of these birds was done largely on the dry banks of prairie ponds where great numbers settled down together. On windy days they sought the shelter of road-banks and other similarly sheltered places.

Other than stubble flights, movement was confined to flights resulting from disturbance and to aerial pre-nuptial courtship.

Sexual behavior.—By the time the puddle ducks reached Delta in April, most of them were paired. On only two occasions have I seen pintail courtship display in the large flocks at rest on the Portage Plains. I suspect, therefore, that the last hens to obtain drakes did so before they reached their breeding marshes. I never have seen mallard displays in spring concentrations at Delta.

Of the time of pairing in the mallard and pintail, Hawkins observed that most of about ten thousand mallards in gatherings near Amarillo, Texas, on January 24, 1943, were in pairs; and that, on the same day, a smaller percentage of pintails were paired (Hochbaum, 1944:21).

Thus, in a spring concentration of loafing pintails and mallards such as those found in Manitoba, few courting birds were seen. One species, the baldpate, was an exception to this rule. When a flock contained some baldpates, the air was full of courting parties. This species seemed to be the last of the puddle ducks to pair. On one occasion in the spring of 1947, I saw 23 drakes pursuing one hen over the Delta marsh. In the spring of 1950 large groups of baldpates stopped on a flooded pasture in the south end of the study area. Courting parties of four to ten drakes were common. These birds were transients as few baldpates nest in the Delta region.

The green-winged teal, an uncommon migrant through Delta and a rare nester in the area, was seen in courtship once. That occurred in the spring of 1949. At that time the paired mallards and pintails were gregarious and tolerant, sitting side by side, sometimes in close groups, with little or no sexual friction.

Distance traveled and time involved.—In regard to the time of departure from the southern marshes Bent (1923:144) quoted Beyer as saying of Louisiana, ". . . winter visitant individuals of the mallard, move northward very early, probably never later than the middle of January." Heit (1948:329) gave a January high for the pintails wintering along the Texas gulf coast and a large drop in February. He said: ". . . the pintails were the first to become restless, leaving in large numbers early in February." Of the northward flight of the pintail in Louisiana, Lynch (letter, January 16, 1948) wrote me, ". . . the returning transients show up here in January (1947, first week in January, 1949, middle January) . . . and the migration lasts for about a month." Hence, by the time the ducks reached Delta, they probably had been on the way for about two and a half months.

Porsild (1943) said that pintails arrived May 9, 1934, in the MacKenzie River delta. Dixon (1943:53) gave May 29 as the arrival date for pintails on the arctic coast of Alaska in 1914. Dice (1920:177) set the pintail's arrival in interior Alaska on May 13. Gabrielson (letter dated January 25, 1951) put the early dates for spring arrival of pintails in Alaska as follows: "Selawik River, April 12; Adak Island, April 13; Forrester Island, April 20; Kuiu Island, April 25; Ketchikan, April 25; St. Michael, April 28; Nulato, about May 1; Bethel, May 1; Kantishna, May 7; Hooper Bay, May 8; Porcupine River at the

141st meridian, May 9; Base of the Kigulik Mountains, May 11; Kobuk River, May 14; and Fairbanks, May 17."

Hence, many of the ducks that stopped at Delta had been on the way for two months since leaving their winter home far in the south. For those whose destination was the far north the entire migration flight would take about three months and the distance traveled would be in excess of 4,000 miles.

Species association during migration.—The species found migrating together most commonly at Delta in spring were the mallard and pintail. As has been shown in Table 2 these species arrived at nearly the same time. Also, the main passage of these ducks through the area occurred at the same time as shown in Table 3. Their course lines, time of day of migration, loafing areas and feeding areas generally were the same.

On the flooded prairie fields and pastures it was not uncommon to find nearly all species of waterfowl together in the spring. The baldpate, which moved through the Delta region in large numbers, was associated with pintails and mallards constantly in loafing concentrations. The blue-winged teal, green-winged teal, shovellers and gadwalls preferred areas where at least some of their food could be taken as it floated on the surface of water. For this reason, areas that dried up in summer and then came up to heavy growth of seed-bearing plants were especially attractive to these species. Such areas were numerous on the Portage Plains and, when flooded in spring, attracted large numbers of migrating ducks from this group.

Flock size during migration.—Only for the mallard and pintail at Delta was it possible to get extensive data on the size of migrating flocks. Sample counts of 54 such flocks of these species have averaged 12 birds per flock. The biggest flock in this sample numbered 35. Occasionally a lone bird was seen migrating along the path taken by the others.

Sex ratio of spring flight.—Hochbaum (1944:15) has shown that a great preponderance of drakes exists among diving ducks as they arrive at Delta in spring. Of the mallard and pintail, he says that an almost exact 50:50 ratio exists as these species arrive. That only a slight excess of drakes exists in these two species is borne out by the sex ratio data collected

by the entire staff of the Delta station since 1939. In a sample
of 3,394 mallards counted between 1939 and 1950 the sex
ratio was 52:48 in favor of males. During the same period, a
sample of 4,926 pintails showed a sex ratio of 52:48 in favor
of males. There appeared to be little variation from one year
to the next, even though yearly samples were small.

SUMMARY

1. Mallards and pintails were the first ducks to arrive on the
northern prairies in spring. For these two species there may be a
yearly variation of about three weeks in time of first arrival. Later-
arriving species such as the shoveller and gadwall showed less
variation in date of spring arrival.

2. The mallards and pintails often migrated together in spring;
most migrating flocks of these birds were made up of both species.

3. From evidence obtained through banding it was found that
resident ducks arrived in the area before the migrants bound for
distant places passed through.

4. Temperature was the dominant factor affecting the northward movement of mallards and pintails but had less effect upon
the migration of the later-migrating species.

5. Reverse migrations were seen occasionally following freezing
weather after the first ducks arrived on the northern marshes.

6. Abortive migration attempts followed long periods of undesirable weather which delayed the ducks in their northward
journey. Ducks were seen in an apparent attempt to migrate but
were turned back by bad weather.

7. The favorite resting places for surface-feeding ducks upon
arrival on the northern prairies were shallow, flooded grain fields.
Their activity there was confined largely to feeding, loafing and
courting.

8. Nearly all mallards and pintails were paired by the time they
arrived in Manitoba. The baldpate was the latest to pair, and
pairing courtship was common during the spring passage of this
species.

9. The northward movement of some species of waterfowl begins as early as January and is not terminated for those going the
farthest until the middle of May. The distance traveled by some
must be in excess of 4,000 miles between winter headquarters and
summer breeding marshes.

10. The sex ratio of mallards and pintails arriving at Delta in
spring was nearly even.

CHAPTER III

MIGRATIONAL HOMING

*T*HE ABILITY of wild birds to "home" has been known for many years. According to Watson and Lashley (1915:9), the earliest mention of the use of homing pigeons is found in the writings of Anacreon (born 550 B.C.). As centuries passed, the use of birds, especially pigeons, to carry messages became important both in peace and war, and consequently is well known to the public.

This early use of birds based on the homing phenomenon has been called "experimental homing" as contrasted with the natural movement of wild birds which has been referred to as "migrational homing" (Hickey 1943:38-49). The number of species of birds in which migrational homing has been found now includes representatives from nearly all families. Thanks to the modern bird band and intensive studies, a vast store of knowledge on migrational homing now exists.

To the ornithologist, the ancient problem of homing still holds many unanswered questions. No completely satisfactory answer has been given to the question of how birds are able to direct their spring flight to the same meadow, to the same limb on the same tree, or to the same small marsh where last year's nest was located. The reverse fall flight to the wintering grounds holds the same mystery; for on this flight, too, there is a return to the wintering areas used the previous year.

Homing is not only an ornithological problem but is of concern to the evolutionist. As Mayr (1942:241-2) has pointed out, the homing phenomenon and social organization in geese probably has been responsible for the development of from six to nine separate geographic species in the genus *Branta*. The tendency for family groups of geese to home to certain restricted areas to breed has resulted in close inbreeding and has caused genetic isolations which prevented gene exchange between related groups.

To the waterfowl manager, the problem of homing is of both intense academic interest and practical importance. It was from the waterfowl manager's viewpoint that experiments at Delta were launched in an attempt to determine the significance of migrational homing to management. An attempt was made to determine where adult hens nested in relationship to where they nested the year before and what percentage of the population returned. The same adult hens, captured on the nest and banded for the renesting study, were used as a basis for these experiments. Also, we tried to compare homing rates with calculated survival rates to determine the percentage of return we logically could expect. Then we attempted to compare this return between adults and juveniles and, also, between the different species of ducks nesting at Delta. Juvenile captive-reared ducks were raised from locally gathered wild eggs and released at Delta. The nest locations of those juvenile hens, and the percentage of hens returned, were studied carefully.

Some of the hens showed a remarkable ability to return each year to nest very close to the site of the nest of the previous year. The record in this study was held by a shoveller hen which returned to the same meadow to nest for four successive years. On June 3, 1947, I trapped and banded this hen on her nest near the center of the study area. The following year I trapped the same hen on a nest 80 yards from her 1947 nest site. Again, in 1949, she returned. Her 1950 site was located just 210 yards from her 1949 nest (Figure 6). Thus, during four consecutive years, she was known to come back to the same nesting meadow within an area of approximately 75 acres. Since no waterfowl winter in Manitoba, and Manitoba shovellers probably go to the Gulf Coast States of the United States or farther south to winter (Kortwright, 1943:-221), the return of this hen involved two flights of at least 2,000 miles each year.

Homing of adult hens.—During the 3-year nest-trapping

period, 152 adult hens of the five species of meadow-nesters were trapped on their nests and banded with aluminum and colored bands. Of this number, 41, or 27 per cent, were known to return and nest in the same meadow a following year. These data are summarized in Table 4.

In order to determine how strong the tendency to return is, we need to know what percentage of the *survivors* returned. Thus, in considering the statistical validity of our data, we must bear in mind two things: (1) the percentage of returning birds that are found, and (2) the expected mortality between one nesting season and the next. To appraise these data, I have treated each species separately.

Although I know that some returning birds were not found, I believe that the number missed was small. The habitat of the study area was such that excellent opportunties for seeing birds existed. Bare, grazed edges of the roadside ditch and of several large potholes made excellent loafing areas where banded hens could be observed through a 20-power spotting scope. Anchored logs put in ditches and sloughs made even more loafing places. In early morning, when the road was untraveled, hens frequently were seen standing on the gravel, and many were identified there. In the spring of 1950, late cold weather froze the shallow ponds, and there mallards and pintails, with banded legs exposed, loafed as a flock on top of the ice. Furthermore, continuous nest-trapping put numerous hens in my hands and gave me an opportunity to see their leg bands.

TABLE 4. NUMBER OF ADULT NEST-TRAPPED HENS KNOWN TO RETURN TO THE SAME NESTING MEADOWS A SECOND YEAR

Species	Banded	Known to Return	Per-centage
Mallard	15	2	13
Pintail	44	17	39
Gadwall	16	6	37
Shoveller	19	8	42
B-W Teal	58	8	14
Totals	152	41	27

Some female birds were missed on the loafing bars; for banded hens not seen previously were nest-trapped and identified. Again, when three were caught wearing bands put on two or more years earlier, and which had not been found during the intervening years, there was evidence that some had been missed. One shoveller banded in 1947 was not found in 1948, nor 1949, but was caught in 1950. Another shoveller hen banded in 1948 was not seen in 1949 but was seen in 1950. No mallard, pintail, or blue-winged teal hens were found returning to the study area a third year that had not been seen the year after banding.

The question may arise: Could some returning hens have nested just outside the study area and been missed? As will be shown in the discussion of nest locations, it is unlikely. Had any nests of returning hens been located outside the study area, their distances from the previous year's nest sites would have had to be at least four times as great as the maximum known distance between the yearly nests of individual birds. This is due to the fact that the study area was surrounded by dense stands of phragmites.

Some clue to expected mortality can be obtained when we examine published data from banding studies. In the appraisal of my homing returns as compared with expected survival, I have chosen to begin with the pintail, the species on which I have obtained the most data.

Adult hen pintails.—During the three years of banding, after which a following year's observation was made, 44 pintail hens were nest-trapped. Of this group, 17, or 39 per cent, were known to return a second year to nest.

No exhaustive work has been done in all of the North American flyways to show how long pintails survive or how many can be expected to return a second or successive years to the same nesting meadows. Furthermore, there is no way of knowing how old the pintail hens were when first banded on my area.

In British Columbia, Munro (1944:61) banded 6,554 pintails between 1923 and 1943 and obtained a recovery of 1,154. Munro says, "Of this total 926 were taken in the first year after banding and the remainder as follows: second year, 140; third year, 36; fourth year, 19; fifth year, 14; sixth year, 6;

seventh year, 5; eighth year, 1; ninth year, 5; eleventh year, 2." Thus, 80.2 per cent of the recoveries were made the first year after banding.

Because this series of recoveries includes birds banded up to 1943, its summarization in 1944 inevitably leaves the older age classes represented unequally. One can, however, study the series as a minimum estimate of the average survival rate of adult pintails on the Pacific Coast during the years between 1924 and 1943. To do this, I have assumed for the purpose of this analysis that all of Munro's recoveries referred to mortality reports (a small minority undoubtedly did not), and I have excluded reports of birds recovered within one year of their date of banding (since many of these unquestionably were banded while juveniles). The method of analysis otherwise follows that of Lack (1940:168), and is summarized in Table 5.

Since additional reports of these British Columbia birds in their eleventh year of life are possible as late as 1954, it is clear that the average survival rate of these pintails probably exceeded 50 per cent per year. Average annual survival rates for mallards and black ducks banded in Illinois already have been calculated as slightly in excess of 50 per cent (Bellrose and Chase 1950:12). In an examination of the mortality rate for adult mallards of unknown age at banding, Hickey (1952:-70) computed a mean mortality rate of 47.7 per cent. The figure here obtained is a plausible one.

If we use 40 per cent as an approximation of the known return of banded pintails to Delta, and assume that another 10 per cent returned but were not found, the homing rate would be 50 per cent. If the survival is in the neighborhood of 50 per cent, this would mean that 100 per cent of the survivors returned. In any event, it is clear that the majority of surviving pintail hens do return to the vicinity of their previous nesting sites.

Adult hen shovellers.—Of 19 adult hen shovellers trapped and banded during 1947, 1948, and 1949, eight were back on the same nesting meadow in years to follow.

Little published information exists on survival rates for this species. At the Bear River Refuge in Utah, Van Den Akker and Wilson (1950:371) say that of 1,087 shovellers banded,

TABLE 5. CALCULATION OF APPROXIMATE SURVIVAL RATE IN PINTAILS FROM DATA GIVEN BY MUNRO (1944:61)

Years After Banding	No. Dying Each Year	No. Alive at start of Each Year	No. Alive at end of Each Year
1-2	140	228	88
2-3	36	88	52
3-4	19	52	33
4-5	14	33	19
5-6	6	19	13
6-7	5	13	8
7-8	1	8	7
8-9	5	7	2
9-10	0	2	2
10-11	2	2	0
Total	228	452	224

Average Minimum Survival Rate $= \dfrac{228}{452} \times 100 =$ about 50 per cent per year.

89 recoveries were reported. (Fifty-five, or 61 per cent of these, occurred during the first year; 22, or 24 per cent, the second year; and the remaining 17 per cent in the third and successive years.)

The calculated crude survival rate of this series is 41 per cent (all cohorts included because most birds were not juveniles when banded), and the true survival rate lies between 31 and 51 per cent.

There is no way to determine how closely these western data apply to Delta shovellers. If we assume, however, that the Delta shovellers have a survival rate similar to the Bear River shovellers, a minimum homing rate of 51 to 100 per cent occurred, and it seems likely that the true rate of homing of adult shoveller hens in Manitoba lies within this confidence belt.

Adult hen gadwalls.—Sixteen adult gadwall hens were nest-trapped and banded during 1947, 1948, and 1949. Of this group six returned to nest again in the same meadow.

Here too, no accurate figure on average survival exists for the Mississippi flyway population. In the west, where it is a more abundant species, Van Den Akker and Wilson (1950:-366) show that of 310 gadwalls banded, 31 returns were received.

The confidence belt of the adult survival rate of this small sample extends from 35 to 69 per cent. The homing rate of adult gadwall hens in the Delta area lies in the range of 54 to 100 per cent, if the survival rate is similar to the gadwalls banded in the Bear River area.

Adult hen blue-winged teal.—The blue-winged teal leads the list in the number of adult nest-trapped hens banded during this study. But despite a comparatively large number of banded birds, the return of hens to the same nesting meadows a second year was small. During 1947, 1948, and 1949, 58 adult blue-winged teal hens were nest-trapped and banded. In the following three years of observation, only eight, or 14 per cent, of these hens were known to return a following year.

The true survival rate lies within the confidence belt of 33 to 53 per cent, and the true rate of homing of adult blue-winged teal hens lies within 26 to 42 per cent.

Considering the consistently higher returns from smaller samples of pintails, shovellers, and gadwalls, I believe that this low return of teal has significance.

For this species, better survival information exists. Bellrose and Chase (1950:14) examined the returns from 6,252 blue-winged teal banded in Illinois and found that the yearly mortality rate for the first five years following banding averaged 57 per cent. In other words, the yearly survival rate would be 43 per cent.

Of all ducks for which calculations have been made, the blue-winged teal has the lowest survival rate. One would therefore expect a relatively lower rate of homing. This may not be the only reason for low homing return. Perhaps the blue-winged teal does not have as strong an instinct to return to the same nesting meadows as does the pintail. On the Delta study area, the nesting population of this species climbed rapidly during the period of study. The climb occurred despite the fact that in the course of the renesting study few nests hatched on the area. Where the birds which made up the population increase came from was not known.

Adult hen mallards.—During the three years of banding, 15 adult mallard hens were nest-trapped. Although this figure was about the same as that for gadwalls and shovellers, only 13 per cent of the mallard hens, or about one-fourth as many as gadwalls and shovellers, were known to return a second year to nest.

This low mallard return may have been caused partly by the difficulty in trapping mallard hens on their nests. Furthermore, the mallard is a relatively mobile species, and the loafing places of the hens which nested on the study area may have been so far away that I missed seeing them.

The relationship between hunting as a mortality factor and survival rates, and in turn also homing rates, is not clear. In summing up the mortality and survival rates of mallards and blue-winged teal, Bellrose and Chase (1950:23) said: "Although the blue-winged teal does not suffer the shooting losses that mallards do, its mortality is higher." These writers found through a study of banding returns that the yearly survival rate in mallards was 50 per cent and only 43 per cent in blue-winged teal.

In terms of the percentage of ducks in the hunters' bag, the mallard rates high throughout the Mississippi Valley. At Delta, during the years 1938 through 1941, they made up 34 per cent of the entire bag (Hochbaum 1944:133). During the years 1946 through 1950, they made up 36 per cent. In the Illinois Valley, Bellrose (1944: 343-5) found that mallards and black ducks together made up between 57 and 75 per cent of the bag between 1938 and 1942.

Summary of adult hen data.—Not less than 50 per cent of the adult pintail, gadwall and shoveller nest-trapped hens of unknown age probably returned a second year to the same nesting meadows. Only 10 to 15 per cent of mallard and 26 to 42 per cent of blue-winged teal hens returned.

A summary of data pertinent to the homing of ducks to their breeding ground is seen in Table 6. A low kill of pintails at Delta coincides with a high return of pintails a second year, and a high kill of mallards at Delta coincides with a low return of mallards a second year. The calculated survival rate for these two species, however, is nearly the same although the issue is clouded by the fact that the pintail survival figure was calculated for western birds instead of Mississippi Valley birds.

A low kill of blue-winged teal at Delta does not coincide with a high return. However, the low return of this species is reflected in the low survival figure. This is in agreement with the conclusion of Bellrose and Chase (p. 25) who say that although the hunting loss in blue-winged teal is lower than the hunting loss in mallards, the mortality from other causes obviously is higher.

It is clear that we need to know more about the survival rates of the pintail, shoveller and gadwall in the Mississippi Valley.

Locations of second-year nests.—Returning hens followed a consistent behavior in locating their second-year nests. On the long narrow study area, all adult hens returned to the same general portion of the area where their former nest or nests had been located. The average, maximum and minimum distances between nests of different years are given in Table 7. The locations of the nests in relation to each other are given in Figure 7. It will be noticed that this map shows a pattern similar to the renesting map in Chapter IX.

Adult hens returning several years.—Table 4 includes hens which returned to nest for two or more years, but does not show the persistency with which some individuals returned. Of the 15 pintail hens banded in 1948, six, or 40 percent,

---

TABLE 6. DELTA KILL, RETURN A SECOND YEAR, AND CALCULATED SURVIVAL RATES FOR FIVE SPECIES OF DUCKS

Species	Place in hunters' bag at Delta 1938-1950	Number of adult hens banded	Adult hens returning	Calculated survival rate
Pintail[1]	4%	44	39%	50%
Gadwall	5%	16	37%	?
Shoveller	4%	19	42%	?
B-w teal[1]	4%	58	14%	43%
Mallard[1]	34%	15	13%	50%

[1] Survival rate of pintail after Munro (1943), blue-winged teal and mallard after Bellrose and Chase (1950).

TABLE 7. DISTANCES IN YARDS BETWEEN NESTS OF
DIFFERENT YEARS OF INDIVIDUALS OF FIVE SPECIES

Species	Number of hens	Maximum Distance	Minimum Distance	Average Distance
Mallard	1	500	500	500
Gadwall	5	1600	120	830
Pintail	9	750	15	315
Shoveller	7	1160	30	306
B-W Teal	5	750	40	253
Total and Averages ...	27	952	141	441

returned in 1949, and three of them, or 20 per cent, returned
again in 1950. Of the shovellers, one hen was known to return
for three consecutive years after first banding; one other,
banded in 1947, was not seen in 1948 or 1949, but was found
nesting in the area in 1950. Of the nine shovellers banded in
1948, three were seen on the area in 1949 and one of them
again in 1950. Of the eight gadwalls banded in 1948, three returned in 1949 and two were back on the area again in 1950.
One of the 14 blue-winged teal hens banded in 1948 was back
on the area again in 1950, although only three of the 1948
birds had been seen there in 1949. No mallards were known
to return for more than one year following their first nest-
banding.

Again the record leads one to conclude that ducks normally
return to their original nesting meadow a second and third and
fourth year or so long as they survive and the area remains
suitable.

Return of juveniles.—Until recently there has been considerable doubt that juveniles return to their natal home. Lincoln
(1939:74) believed, that though there was a high juvenile
mortality, evidence from the banding data appeared to indicate that "homing instinct" did not operate until after the individual had nested and that the location of the first nest was
largely a matter of chance in obtaining possession of unoccu-
pied territory.

Nice (1937:185) refutes Lincoln's theories, and concludes
that: "Young are certainly far less *ortstreu* or faithful to their

Figure 6. Shoveller hen No. 47-604004 returned to the same nesting meadow in 1947, 1948, 1949 and 1950.

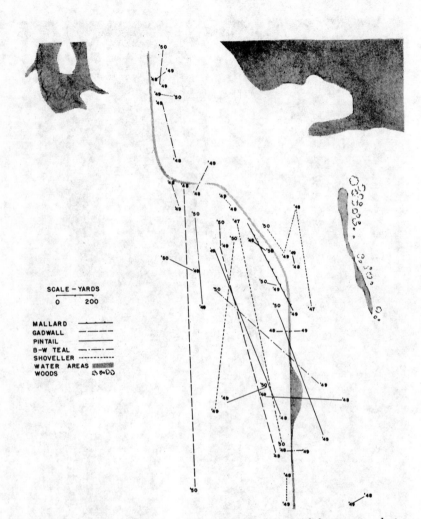

Figure 7. Nest locations of returning individual hens of five species during different years.

home than are adults. Young of many species do scatter widely, although I never will believe that they scatter over the
whole 'natural range of the species,' or even subspecies. In
some cases a substantial proportion of the surviving young
has been found to return to the vicinity of the birthplace."
Nice cites numerous instances of young returning to their
birthplace the second year to substantiate her statements.

That this return is not a random dispersal has been shown
by Farner (1945:97) who says of the eastern robin: "There
is a marked tendency among robins, *Turdus migratorius migratorius*, to return, as breeding birds, to their birthplace or its
immediate vicinity. . . . There is no random dispersal among
first year birds."

What of returning juveniles among ducks? Valikangas
(1933) found, that mallards, hatched in Finland from eggs
transported from England, returned to Finland to nest the
year following their release. In this instance, the eggs were
from resident non-migratory birds. The return the year following release was extremely high, being between 55 and 58
per cent of the number released. McCabe (1947:107) conducted a similar experiment with wood ducks in Wisconsin.
He transplanted young wood ducks from Illinois, reared and
released them in Wisconsin, and obtained circumstantial evidence that these wood ducks returned to nest in the same area
where they were released.

In an attempt to measure the return of juvenile birds to
their natal pond, we began a program of releasing young birds
at Delta in 1948. During 1948 and 1949, 370 captive-reared
juveniles, raised from local eggs, of five species were released
from the Delta hatchery. These birds were color-banded so
that they could be identified easily if they returned to the area
the following year. They were released at the age of five to
six weeks, and they grew to flying age in the hatchery pond
where they mingled and finally migrated with wild birds.

Because the pintail was the most abundant of the early
meadow-nesting species, and the most easily raised in captivity, it was the species most emphasized in this work. It
was a logical choice, also, because more data on wild hens
of this species were available for comparison with captive-
reared juveniles.

In addition to the release of captive-reared birds, I trapped and banded wild birds in the fall on the study area. As these birds were sexed and aged at the time of capture, the juvenile hens among them gave further data for this discussion. Since their place of hatching was unknown, I will refer to them as of unknown origin.

The data concerning released captive-reared juvenile birds are given in Table 8. Comparisons of the number of returning captive-reared juveniles, with the number of returning adult nest-trapped hens, are given in Table 9. The species order in which juvenile captive-reared hens returned followed closely the order in which adult nest-trapped hens returned. Thus, pintails, which were banded in greatest numbers both as adults and captive-reared juveniles, showed the highest return in both categories. Mallards and blue-winged teal showed a low return in both categories. This seems significant, and I believe we can conclude that the tendency to return is greater in the pintail than in the mallard and blue-winged teal.

I already have discussed the percentage of adult birds that returned to the same nesting meadows the year following banding, and speculated on the percentage of survivors we could expect to return. As noted in Table 9, the percentage of juveniles returning to the area where they were raised was much lower than the return of adult birds. Three factors may contribute to this difference: (1) juvenile birds may not be as exacting in their homing as adults; (2) a lower percentage

TABLE 8. NUMBER OF CAPTIVE-REARED JUVENILES KNOWN TO RETURN TO AREA OF RELEASE THE FOLLOWING YEAR

Species	Males		Females	
	Released	Returned	Released	Returned
Mallard	13	0	20	1
Pintail	132	2	115	15
Gadwall	9	0	8	1
Shoveller	12	0	12	1
B-W Teal	19	0	30	0
Totals	185	2	185	18

TABLE 9. COMPARISON OF RETURN OF ADULT AND JUVENILE HENS OF FIVE SPECIES

Species	Adults		Juveniles	
	Number Banded	Percentage Returned	Number Banded	Percentage Returned
Mallard	15	13	20	5
Pintail	44	39	115	13
Gadwall	16	37	8	12
Shoveller	19	42	12	8
B-W Teal	58	14	30	0
	—	—	—	—
Totals	152	185		

of juveniles than adults probably survive to return; (3) since these juveniles were captive-reared their survival rate may not be so great as that of wild birds.

The area to which adult hens returned was near the area where they nested the year before. The knowledge or memory of an area probably is stronger among adults than among juveniles because of the greater length of time that adults spend on a particular area. Thus, it seems plausible that more adults would pinpoint their return to a smaller area than juveniles. There may be another reason for the higher and more exacting homing ability of adults. Juveniles may arrive on the breeding marsh later in the spring after the adult birds already have set up their territories. In the spring of 1949, when information was obtained on the return of the early pintail hens to the place where they nested the year before, it was noted that the adult resident birds were the first to arrive at Delta and that they preceded the main flight of transients. Adult pintail hens returning to the same meadows where they nested the year before appeared at Delta before the returning juveniles that had been raised and releasd there the previous year. If adults arrive on the breeding ground before the juveniles, they have a definite advantage in competing for space. This may spread the returning juveniles over a larger area than the returning adults, which nest close to the place where they nested the year before.

It is well known that juvenile birds have a much lower

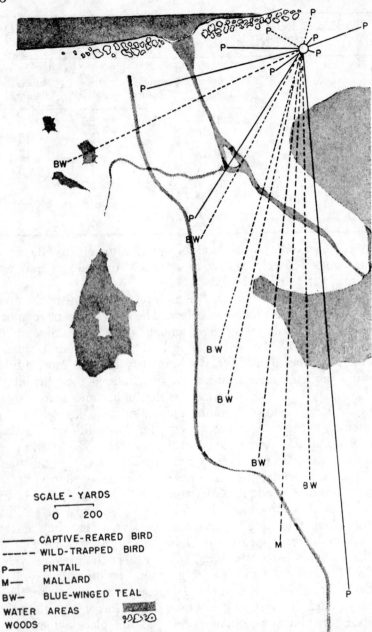

Figure 8. Relationship of summer release point of captive-reared juvenile hens and wild-trapped juvenile hens to nest site the following spring.

survival rate than adults; and, because of the heavier mortality among juveniles, we would expect the percentage of homing juveniles to be lower than that of homing adult hens.

If we compare the return of juveniles with the return of adult birds banded at Delta, the lower return of juveniles may be attributed partly to the fact that the juveniles released were captive-reared rather than wild birds. In an analysis of banding returns, Hickey (1952:69) compared the survival of hand-reared birds with the survival of wild birds and concluded: "Samples of hand-reared birds were examined in passing. Among these, the percentage shot during the first year of life did not vary with the month of banding (June, July or August). It did, however, markedly differ from that for wild-reared birds. Preliminary data also indicate that the adult survival rate of wild birds may not be attained by these hand-reared individuals until the third year of life. By this time, however, only 5 per cent of those reported are still alive."

Nesting of returning juvenile hens.—Of the 18 captive-reared juvenile hens which were known to return to the hatchery pond the year following release, nests of eight were found. The locations of these nests, as well as the locations of nests of nine returning wild-trapped juveniles, are shown in Figure 8.

The distances from the autumn release point to the spring nest sites of these eight hens (all pintails) were 40 yards, 40 yards, 190 yards, 300 yards, 400 yards, 800 yards, 1,000 yards and 2,600 yards.

Two wild juvenile pintail hens of unknown origin, which were trapped and banded during the fall, nested the following spring 180 yards and 110 yards from the point of capture. One juvenile mallard hen, captured in the hatchery pond and released in September 1948, nested 2,500 yards from this pond the following spring. Six juvenile blue-winged teal hens of unknown origin, captured and banded in the hatchery pond, nested on the study area the spring following their capture. For these hens the distances between the fall capture point and their spring nest sites were 1,080 yards, 1,375 yards, 1,550 yards, 1,700 yards, 2,000 yards and 2,200 yards.

The dispersal of juvenile hens for their first nesting is a difficult thing to determine. How many juvenile hens survived and nested outside the study area I do not know. I believe there is some significance in the following facts. During the years 1946 through 1949, I trapped 693 wild juvenile hens of unknown origin at the hatchery pond, and nine of these hens were known to nest the year following within 2,500 yards of the point of capture and release. During the same period, Arthur S. Hawkins captured, banded, and released 1,947 (almost three times as many) juvenile hens at two points, one mile and three miles southeast of my banding station; but none of the birds banded by Hawkins were found nesting on my study area. No check was made to determine how many of the birds trapped by Hawkins returned to nest near the place where they were banded.

Return of juvenile drakes.—From Table 8 we have seen that of 185 juvenile females banded, 18 returned to the same area the following year. Of the same number of males banded, only two were known to return a second year. When these drakes were seen, neither was accompanied by a hen.

It generally is assumed that waterfowl pairing takes place on the wintering ground, and that the hen brings back to the nesting place a drake which probably had no previous experience in that particular part of the breeding area. However, if males and females hatched and raised near each other migrated together and wintered together as members of a unit population, it would be reasonable to expect that a male might return to the vicinity of its natal marsh in company with a hen also raised in that vicinity.

In the spring of 1950 I collected a mallard drake at Delta that had been banded by a co-worker three miles southeast of my study area the previous fall. This drake was with a hen at the time he was collected. It seemed likely that this male had returned with a hen to nest in her natal marsh which was in the vicinity of either his natal marsh or a fall loafing area. That males, as well as females, probably migrate through the same areas year after year has been indicated by Munro (1943:235) and Cartwright (1945:337).

The probability is less than one in 500 that the differences in the rate of return to the area of release, shown between juvenile males and females in Table 8, could have occurred purely by chance. There is little doubt that it is the female that pin-points her flight to a definite area in spring and that the male, although capable of doing so, follows the hen to the place of her return and only coincidentally arrives at the same place he was the year before. This is in agreement with the conclusions made by Lincoln (1939:75) who says that the evidence now available indicates that among ducks it is the influence of the female that determines the movements of the male.

Homing and waterfowl management.—From the foregoing data it is clear that most surviving adult hens of at least some species return to the marsh where they nested the previous year. It is clear too, that a smaller percentage of the surviving juvenile hens also return.

Of what importance is this to waterfowl management? Hochbaum (1947:56) has advanced the theory that local breeding populations are vulnerable to annihilation when the ducks that normally would home to an area are hunted too heavily on that area. He calls this homing tendency "breeding tradition" and says that many good marshes are devoid of breeding birds because of the loss of breeding populations owing to early and heavy hunting. Hochbaum (p. 55) calls the marshes that have lost their breeding populations, "burned out" marshes, and says that the reasons for burning out are: (1) cropping too early and (2) harvesting too heavily. As the season progresses, migrant birds mingle with the resident birds and gunning pressure is spread more thinly over many units of population rather than being concentrated on one.

The effects of later hunting seasons have not been measured thoroughly; but in a few instances there are indications that later seasons may be bringing back duck populations. In the Delta marsh, the hunting season in 1946 opened on September 16. In 1947, and the years immediately following, the season did not open until the second week of October. During the years between 1947 and 1950, substantial increases in breeding populations of blue-winged teal and redheads were noted. In 1946 an estimated 15 to 20 pairs of blue-winged teal were breeding on the study area; in 1947, 25 to 30; in 1948, 30 to 40 pairs; in 1949, 45 to 50 pairs; and in 1950, 50 to 60 pairs. Being an early-migrating species in autumn, the blue-winged teal has escaped much hunting pressure at Delta during the last four years mentioned when later openings were the rule.

The ability of ducks to repopulate areas where the breeding population has been lost, or to populate new areas, depends largely upon their ability to pioneer into those new areas. Our homing data would seem to indicate that the juveniles are more likely to populate new areas than adults. Hochbaum (1946) believes that the puddle ducks may be more inclined to pioneer successfully than are the diving ducks and says (p. 407-408):

". . . when new water areas are created there is a response on the part of certain ducks which come to nest at these new places. We see this in the new refuge marshes. It was shown by the 1945 behavior of river ducks on Manitoba farmland. These birds, in their many thousands, bred on new waters which had not existed within their individual life spans. In other words, these ducks moved to areas with which they could have had no previous experience and which had not been used by ducks for at least one waterfowl generation. This is pioneering. Through its ability to pioneer a species responds rapidly to management. If it pioneers slowly, it responds slowly to management.

"Clearly the most successful river ducks are the most rapid pioneers. The diving ducks pioneer slowly. River ducks being more tolerant in their choice of breeding environment than diving ducks find a wider variety of ecological patterns acceptable. . . .

"I suspect, however, that variation in the pioneering trait reflects more than variations in breeding tolerances. The ability to pioneer is a part of the specific makeup, and there is much variation species by species. The ability to pioneer is highly developed in the mallard and pintail, poorly developed in the redhead and canvas-back."

The Delta homing data points to the conclusion that juveniles pioneer to new areas more readily than do adults. If the adults arrive first on the breeding range in spring, as is indicated (Chapter II) by the return of banded pintail hens, the juveniles are subject to internal population pressures that may force them into new nesting areas. At no time during these studies, however, did it appear that populations had reached the point where internal population pressures were active.

Hochbaum (1946:407-408) describes the pioneering of

large numbers of surface-feeding ducks into new areas when there was no apparent reduction in waterfowl range. In another instance, (Bue, et al. 1952) newly developed waterfowl areas in South Dakota were populated rapidly by surface-feeding ducks. In reviewing this paper Hickey (1953:225) points out, "Ducks invading the cattle-range country of South Dakota were common or fairly common surface-feeding species that were subject to significant shrinkage of habitat through drainage operations immediately to the east."

Thus we see that pioneering has been observed both when there has been and has not been apparent waterfowl range reduction.

The artificial stocking of marshes to build up breeding populations of Canada geese has been successful in South Dakota, Michigan, and Manitoba. For most species of ducks, however, the rebuilding of breeding populations still is in the experimental stages.

Regarding the advisability of using hand-reared birds to restock areas, Brakhage (1953:475) says, "If there is practical value to be derived from the release of hand-reared stock other than satisfying the local gunner on an immediate basis, it must be realized through the return of birds to nest. If the surviving individuals return to their marshes of liberation to nest in the following years, and thus replace a 'burned out' breeding population or colonize a new area, a long term value will be realized. But it cannot be considered sound management to stock birds which do not survive long enough to reproduce." In any attempt to repopulate marshes by liberating birds, Brakhage (p. 476) believes the use of hand-reared birds from wild eggs is better than the liberation of game-farm birds. However, owing to the deteriorating effects on populations in marshes where wild eggs are gathered and to the expenditures necessary to promote an artificial stocking program, he does not recommend the release of ducks as a practical management technique.

Among the unanswered questions regarding the significance of homing are these: What happens to the ducks when whole populations of wide areas are forced to pioneer? Are they as successful in the unfamiliar area? Does a slowly drying marsh invite homing ducks in spring but become inadequate during the critical nesting season? Burning out of marshes can be avoided by later and lighter shooting; but how can a reasonable harvest of early-departing species of ducks be obtained without hurting those that migrate late?

SUMMARY

1. The annual return of wild birds to the area in which they nested the year before, or were raised, has been called *migrational homing*.

2. The existence of migrational homing in wild birds has been known for many years; the percentage of survivors returning, and the differences in return between adult and juveniles, are treated in this chapter.

3. During the years 1947-1949, 152 adult hens of five species of meadow-nesters were trapped on their nests and banded to determine the percentage of return in following years.

4. Of this number, 41 or 27 per cent, were known to return and nest in the same meadow a following year. Mallards showed a 13 per cent return, pintails a 39 per cent return, gadwalls 37 per cent, shovellers 42 per cent, and blue-winged teal a 14 per cent return.

5. When we compared the above return with expected survival it was determined that the majority and probably all surviving pintail, gadwall and shoveller hens did return to the same meadow a second year.

6. Blue-winged teal and mallards did not show as clear a picture of return. Although more adult blue-winged teal hens were banded than any other species, their return was one of the lowest. This may have been due to the fact that the blue-winged teal had a lower calculated survival rate than that of any other species. The low mallard return may have been due to the smaller sample size, and to the fact that mallards were more difficult to work with and to observe.

7. Juveniles, too, showed a strong tendency to return; but the percentage of juveniles returning was far below that of returning adults. Of 370 captive-reared juveniles released during 1948 and 1949, two drakes and 18 hens were known to return the following year.

8. The fact that so few drakes, both juvenile and adult, returned to the same marsh the year following banding probably is due to the fact that pairing takes place on the wintering ground

and that the hen determines the place in which the pair will breed the following year.

9. The theory has been advanced by Hochbaum that local breeding populations are vulnerable to annihilation when the ducks that normally would return to that locality are hunted too heavily.

10. To avoid this annihilation, Hochbaum has suggested later seasons and lighter harvests on breeding marshes.

11. From 1947 to 1949, the opening date for waterfowl shooting in the Delta marsh area was about two weeks later than in previous years. Increased breeding populations of blue-winged teal, pintails and redheads were built up at Delta, but may not be wholly the result of later seasons.

12. The ability of ducks to move to and repopulate "burned-out" areas, or new areas, is called pioneering. The more successful surface-breeding species (mallard and pintail) have shown the greatest ability to pioneer.

13. The repopulation of marshes by artificial release of birds is too costly to be of practical value over large areas.

14. Among the unanswered questions regarding the significance of homing are these: What happens to the ducks when populations of wide areas are forced to pioneer? Are they successful in a new, unfamiliar area? How can a reasonable harvest of the early-departing species of ducks be obtained without hurting those that migrate late?

CHAPTER IV

HOME RANGE AND TERRITORIALITY

*T*HERE ARE few patterns of behavior that are followed so universally by the animal world as the adherence to a home range. The crayfish, the Johnny Darter, the skink, the chickadee and the elk do not wander about this world at random, but each individual of each kind has a special part of the terrain which is its home, its familiar range in which it passes most of its lifetime.

It was Seton (1929) who told us so much about the home ranges of the mammals, who showed how even in the wanderings of the wolf and the mink there is a definite pattern of travel by which the animal confines its activities to a region with which it becomes familiar.

We know from the studies of banded individuals that birds have home ranges in which their activities are more or less limited. This is defined sharply in some sedentary species such as the chickadee, individuals of which may not travel more than a mile from their birthplace during a lifetime. In many other kinds, of course, there is a migration that separates the wintering grounds from the summer quarters of an individual. The home range does not contain all of the area between the two places, but both in summer and winter we find the travels of the birds restricted. In some highly mobile species, like the waterfowl, there may be no home range in the strict sense of the words on the wintering ground; but in the north, in the spring, when the birds settle down to nest, they occupy a definitely limited area.

Within this home range there exists, among other constituents, a *territory*. The behavior of the birds concerning it is *territoriality*.

In order to clarify the relationship between home range and territory in ducks, we need to define these concepts further and discuss them in terms of specific individuals.

In the home range of waterfowl, then, I propose that we include all of the familiar area used by a breeding bird; the defended territory, the nest site, favored loafing spots, feeding places, and all the meadows or intervening spaces frequented by the bird. It is the *area within which a bird spends its period of isolation between the break-up of spring gregariousness following spring arrival and the reformation of fall gregariousness.* It should be remembered that short periods of reversion to gregarious behavior may interrupt this period when "thwarted" hens which have lost their nests join bands of loafing waterfowl between nesting attempts. Home range in ducks is the area within which a bird remains from the time it settles down to breed until the time breeding is over. It is the area with which the bird is most familiar. It is the area to which the hen, so long as she is alive, and so long as it is acceptable, will return year after year to breed.

An example of home range.–Shoveller hen #47-604004, which was known to return to the Delta study area for four consecutive years (at which time the present study was completed), was known to nest each year within an area of approximately 75 acres (producing four first nests and three renests). The home range of this bird is shown in Figure 9. Her home range included an 800-yard segment of roadside ditch, a large slough 450 yards to the east, another slough 400 yards to the northeast, and two small bays. Also included in it were the nesting meadows between and around these water areas. The map of this particular bird's home range would include about 200 acres of land. It was, of course, used simultaneously by many other birds, including shovellers.

In the summer of 1947, the first year that this bird was known to be on the study area, she apparently remained somewhere within this block of land for 93 days.

The chronology of the known events in her life is shown in Table 10.

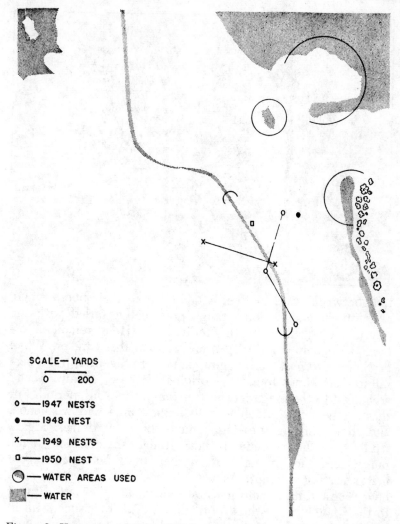

Figure 9. Home range of shoveller hen No. 47-604004 during 1947, 1948, 1949 and 1950.

This record includes the laying data on seven nests (four first nests and three renests) over a period of four years. It represents a total of 65 eggs covering a laying period of 65 days. For the four years that this hen returned to the study area her home range remained the same.

Duration of the home range attachment.—There was varia-

tion between sexes in their attachment to their home range, and attachment to home range presumably varied with the physiological status of the individuals. Hens remained attached to their home ranges longer than their drakes. When first efforts to nest were begun, pairs were together; but a nest failure sometimes found the original drake of a pair already advanced to the gregariousness of the post-nesting period. For these drakes the attachment to home range had weakened, and they joined a growing band of bachelor drakes. Meanwhile their hens retained their attachment to their home ranges and each brought in another drake before making a first renesting attempt. As we will see in Chapter IX, such hens renested near their first nest failure; and their attachment to their home ranges held until they were either successful in hatching a brood or until they ceased to try—a period of sometimes up to three months or longer.

Territory.—The study of the breeding behavior of birds has been advanced greatly by the observations of Howard (1920) in respect to territory, which has been defined by Nice (1937), (1941), Mayr (1935), Tinbergen (1939) and others as a *defended area*. Nice (1941:441-487) discusses the history of the study and gives the following definition: "The theory of territory in bird life is briefly this: that pairs are spaced through

TABLE 10. CHRONOLOGY OF EVENTS IN THE LIFE OF
SHOVELLER HEN #47-604004

1947

May 22	Laid first egg of first clutch of 12 eggs.	During this period the hen frequented a stretch of ditch 800 y a r d s long, one slough 450 yards east of the ditch, a slough 400 yards northeast of the ditch and two small bays 600 yards n o r t h e a s t of the ditch.
June 6	First nest was destroyed.*	
June 17	Began laying second clutch which contained 8 eggs.	
June 27	Second clutch of eggs was destroyed.*	
June 30	Began laying third clutch of 8 eggs.	
Aug. 1	Third nest of season hatched.	

1948

May 21	First seen on the study area with a drake at pole 21. This day she began laying.	Same home range inhabited as in 1947.
June 1	Clutch of 12 eggs completed.	
June 13	Nest was destroyed.*	
June 19	Drake still accompanied hen on her off-nest periods.	

1949

May 9	First seen on study area.	Same home range inhabited as in 1947 and 1948.
May 18	Began laying. Clutch of 1 egg, destroyed same day.*	
May 19	Moved 380 yards to the west and began laying clutch of 12 eggs. Completed laying on June 29.	
June 5	Nest destroyed.*	

1950

May 16	Nest started. Clutch of 12 eggs completed on May 27.	Same home range inhabited as in 1947, 1948 and 1949.
May 30	Nest destroyed by skunk.	
1951	Hen not found on study area.	

---

* Nest destruction here refers to the robbing of eggs as part of the study.

the pugnacity of the males towards others of their own species and sex; that song and display of plumage and other signals are a warning to other males and an invitation to a female; that males fight primarily for territory and not over mates; that the owner of a territory is nearly invincible in his territory; and finally that birds which fail to obtain territory form a reserve supply from which replacements come in case of death of owners of territories."

Most territorial study to date has been concerned with passerine birds. Among waterfowl, too, one is aware of this territorial behavior during the breeding season.

One of the earliest records of such behavior in waterfowl is cited by Nice (1941:442) from Ticehurst who quotes: "Section 21 of 'The orders Lawes and Ancient Customes of Swanns', printed by order of John Witherings in 1632: And yet neither the Master of the Game, nor any Gamster may take away any swanne which is in broode with any other mans, or which is coupled, and hath a walke, without the other's consent, for breaking the brood."

According to Delacour and Mayr (1945:50), Geyr was the first writer to point out that pursuit flights in waterfowl actually were territorial defense flights. Hochbaum (1944:56-78) has given us the most complete discussion of territory among ducks.

The main points of Hochbaum's discussion are as follows: (1) "At the time the pair is ready to nest it takes title to a small water area of the breeding marsh—a pothole, the corner of a slough, or a portion of bay edge. Day after day, as long as the drake and hen remain together as a pair, they may be found on this water area." (2) "The water area occupied by a pair of nesting ducks is defended by the drake; he establishes definite boundaries against the intrusion of other sexually active birds of his own species." (3) "A territory is a specialized piece of terrain in which four components must exist together: Water, loafing spot, nesting cover (adjacent or nearby) and food."

In studies of aggressiveness in birds so much emphasis has been placed upon territory that we have somehow, many of us, come to think of a bird as being encapsulated within this defended territory. This is not true, as Lack (1943) demon-

strated in his study of the English robin. Kendeigh (1947:73)
also demonstrates it in his two studies of territorial behavior
in warblers and says, ". . . males of several species were found
wandering considerable distances outside of their defended
territories." And of the later study he says, "The same tendency
toward a wide home range was noted. . . ." Hochbaum (1944)
too, in his discussion of abandonment of territory (p. 80) and
departures from territory (p. 81) makes qualifications to take
care of movement outside the defended area for breeding
waterfowl.

At Delta, I found that the concept of territoriality as defined above could not always be applied to the birds under
my intensive observation and that, although fitting the general
pattern, their behavior deviated from it in many instances.
Pairs did not always establish definite boundaries to defended
areas. A wide variation between the behavior of individuals
was noted. Some drakes defended from more than one area.
The various components required in a home range did not
always exist inside a defended area.

Territoriality in shoveller #47-604004.—This hen, the
home range of which has already been discussed, was known

to favor a section of roadside ditch bank between poles 15 and 29. She and her drake often were seen there. Although the defensive behavior of her drake was observed there more frequently than elsewhere, it also was observed along other parts of an 800-yard stretch of ditch.

During the incubation period after her drake had left her, this hen returned to the same places on the ditch and slough to feed and preen during her off-nest periods, places which she and her drake had used as loafing spots when they were together.

In order to check the predictability with which I could expect to find this hen on her most favored area (territory), I analyzed all shoveller sight records made at random times of the day, over the four-year period on the section of ditch bank included between poles 15 and 29 during the height of the nesting season. I found that the shoveller population on this part of the ditch varied from zero to seven. On 148 visits to the area during the nesting period, the following numbers of shovellers were seen: none was seen on 36 per cent of the visits; one was seen on 26 per cent of the visits; two on 22 per cent of the visits; three on 4 per cent of the visits; four on 8 per cent of the visits; five on 2 per cent of the visits; six on 1 per cent of the visits; and seven were seen on 1 per cent of the visits. Even during this nesting period when territorial occupancy was most pronounced, hen #47-604004 was not seen on over one-third of my visits to her favored area. Although seven or more shovellers used the area, there was wide variation in the population size at any one time. It is noteworthy that the shoveller, among the species with which we are here concerned, adhered most faithfully to territory and could therefore be said to be the most territorial of the species using the study area. It also is noteworthy that among these species the shoveller drakes remained longest with their hens.

Daily variation in ditch population.—Similar counts for the entire ditch were made which showed a drop in the population from morning to evening. Data from 118 counts taken at various times of the day are given in Figure 10. Lone hens have been omitted from these numbers but are included in the total. Lone hens whose drakes had left them were omitted because they did not indicate territorial occupancy of a breed-

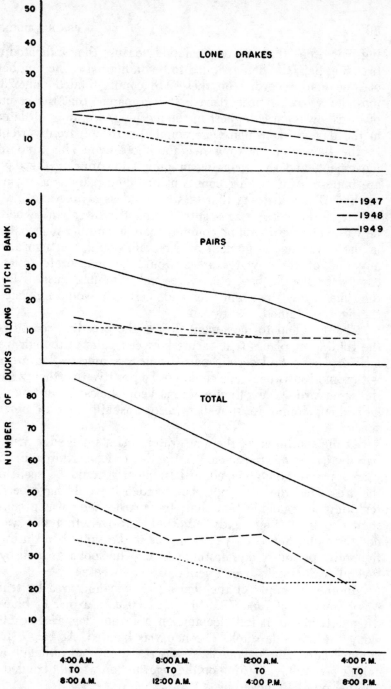

Figure 10. Change in roadside ditch population during the day based on 118 counts made in 1947, 1948 and 1949.

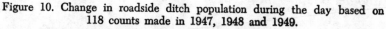

ing pair. Lone drakes were included because they indicated a breeding pair, the hen assumed to be on her nest. The number of lone hens recorded on these 118 counts totaled only 23, most hens having been paired. It is apparent that the count of birds always was highest in the morning, dropping off later in the day. Birds which were counted early apparently went to the bay later, or to a neighboring slough. This kind of movement, and the changing of populations during the day, are important in making counts of breeding pairs as a census method. Thus, a drake that was counted as a breeding individual of the nesting area surrounding the ditchbank where he was seen might well be counted again in another place later in the day. The assignment of a certain population figure to any one of these water areas would be inaccurate unless counts for slough, bay, etc., were made simultaneously. Also, the time of day the count was made certainly would influence the figure obtained.

Closely allied to movement during the breeding season is the trading of places that occurs between pairs. Although we may expect to find certain pairs using the same loafing areas with some consistency, we could not be positive that they were the same birds unless they were marked. The same shifting of individuals found in shovellers also was apparent in other species.

At the south end of the long water area a small edge made bare by cattle was particularly attractive to loafing birds. Each year a pair of blue-winged teal frequently could be seen in the area. Had the birds not been banded it would have been tempting to conclude that the pair (or the hen) was a constant occupant. Leg bands indicated, however, that this was not the same hen each year nor was it the same hen during the entire period of one nesting season. The spot was used by several distinct pairs over a period of a few weeks.

Another example of the sharing of a loafing area by teal was observed on June 7, 1949. At 11:00 a.m. a pair of blue-winged teal was noted sleeping on a loafing log at pole 30 along the roadside ditch. The hen was banded. At 11:45, 14 minutes later, the loafing place was revisited; and although a pair of teal still was sitting on the log, the hen was not banded. Another pair had moved in.

Thus we see that although the birds do use favored spots consistently, they may favor more than one spot; and that two or more individuals or pairs may *share* a certain place, using it at different times.

Not only did the trading of places occur among the more sedentary blue-winged teal, but it also occurred among the mobile species. Close watch of a particular section of roadside ditch was made in 1948. It was used by four pairs of gadwalls where casual observation would have assigned only one pair. The four pairs all came to that part of the study area from distant bays and sloughs and used that portion of the ditch as a jumping-off place on their way to nest sites. The drake in some instances returned to that place to wait for his laying hen and from there sometimes made defense flights. Figure 11 shows the location of this jumping-off place in relation to the nesting places of these birds and the hour and date the individual nests were first indicated as the hens were seen to go to them. Also shown in Figure 11 are the locations of three other gadwall nests found in the same general area during the period, but for which waiting or jumping-off places were unknown.

Duration of aggressive behavior.—The peak of defense flights at Delta was difficult to determine accurately because the flights gave little indication of the number of drakes making them. That is, active drakes took off on defense flights repeatedly during an interval of time; and, as the pair moved about, defense flights by the same drake were initiated from different places.

The actual first peak of nesting was clouded also because there was no certain way of telling which nests actually were first nests and which were early renests (see Chapter IX).

However, taking these difficulties into consideration, there appeared to be close correlation between such flights and the beginning of nesting.

In 1949, defense flights of pintails were at their peak between April 15 and May 5. Occasional defense flights were seen until May 20. During that year the peak of early laying was between April 15 and April 28.

For the same year, the mallard defense flights were common beginning April 10, five days before the first known egg

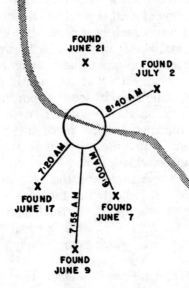

Figure 11. Shared jumping off place and nest locations of four gadwall hens during the spring of 1948 and nest locations of three other gadwall hens.

was laid, and continued to be intense until April 29. They were practically over by May 4.

In 1950, a much later spring, the period of first mallard nests reached its peak between April 29 and May 12. Mallard defense flights were common during the period between April 21 and May 16. The peak of first nesting pintails for 1950 was reached between April 29 and May 12, and the most intense defense flights for this species occurred during the period from April 20 to May 8.

The relationship between the duration of aggressive behavior and nesting chronology at Delta was less pronounced in the gadwall, shoveller and blue-winged teal. This can be attributed to the comparatively small population of the first two species and the less conspicuous defense action of the last species. The small and less mobile teal made only short defense flights along the confines of the ditches.

Absence of aggressive behavior.—Although the territory generally is defined as the defended area, and it is assumed that there is no territory unless it is defended, pairs and individuals continue to use a favored spot after aggressiveness wanes. For this reason we sometimes find the word *territory* used without reference to defense. For example, Munro (1943:239) says of the mallard, "The term 'territory' is used to identify the pond, or portion of stream, or area or lake shore usually occupied by a breeding pair. It is their feeding, breeding and resting place; no behavior that might be interpreted as territory defense has been observed."

Munro (1944:76) found little territorial defense among pintails in British Columbia and says, "There does not appear to be much competition for territories—*i.e.* the pond, slough, or portion of lake shore, usually some distance from the nest, that is occupied by a nesting pair. Where several pairs use the same slough or section of lake shore no hostility between the males has been observed."

Situations closely resembling that described by Munro for pintails were noted at Delta in 1950. During that year, pairs of pintails, including seven hens banded the year before, returned to the natal pond where they had grown up. This group of hens and their drakes used the pond area as a feeding and nesting place during nesting season with practically no evi-

dence of aggressive behavior between them. Nests of six of
these hens were found within 200 yards of the pond. In addition to the pond area which these pairs used intermittently
during the nesting season, the same pairs frequented several
other areas in the vicinity of the research station. At the close
of the nesting season in July, six of these hens remained in the
pond area without drakes and fed, loafed, and remained together as a group for part of the summer before departing for
some unknown molting area.

Variations in territorial behavior.—That territorial behavior
does not always exist among breeding ducks is suggested by
the observations of Munro and my own work at Delta. On the
other hand, evidence of strict adherence to territories in black
ducks has been described by Trautman (unpub. ms. 1949)
who studied a low population of these birds on a lake in Ohio.
Trautman's birds definitely adhered to bounded areas and
were nearly always within them in contrast to the larger movements and overlapping of movements that I observed at Delta.
An unanswered question on the variation one might expect in
territoriality has been asked by Trautman who says, "It would
be most interesting to know if territorialism breaks down,
partly or completely, where there is a greater nesting population, hence more competition and/or where changing water
levels or other factors make territory hunting more difficult."

Differences in territoriality because of population density,
topography, and the interspersion of land-water types with its
resultant pattern of waterfowl requirements, can be studied
only by marking large numbers of drakes and hens during the
breeding season.

Reactions of drakes to "dummies."—The use of mounted
birds to study behavior has been a common technique of ornithologists. Beginning in 1947, I atttempted to use this technique
to test the aggressiveness of males. Mounted specimens of mallards, pintails, shovellers and blue-winged teal drakes in nuptial plumage were prepared in upright standing position for
this purpose. These were placed on loafing sites of drakes
believed to be on territories, and the behavior of each drake
was noted when he returned.

Although numerous tests were made with pintail dummies
at various stages of the nesting cycle of pairs, no response was

elicited from the drakes of this species. On one occasion, a mallard drake would not return to his loafing place on a muskrat house where a stuffed mallard drake had been placed. Shoveller drakes were frightened away from their loafing places by the dummy, although they were observed to bob their heads in threatening postures from a distance.

Wide variations in the reactions of blue-winged teal drakes to stuffed dummies was noted. For example, on May 30, 1948, a stuffed drake teal was placed along the bank where a pair of teal had been resting. The pair flew away when the dummy was placed but returned in about five minutes. This pair swam slowly toward the dummy, the drake leading his hen. When the stuffed specimen was reached, it was attacked violently by the drake. After the dummy had been knocked over, I walked in and placed it in an upright position again in order to save it from destruction. As I did so the attacking drake and his hen flew away. Another drake blue-winged teal, thirty feet farther down the ditch, swam slowly into the first drake's area. Upon sighting the stuffed specimen, the latter drake made two cautious advances toward it and then withdrew to swim back to the area from which it had come. In this particular instance it appeared that the first drake exhibited complete self-confidence and was extremely aggressive on his own territory, while the drake from the adjoining area exhibited only mild aggressiveness when confronting a dummy drake near his territory.

A reaction unlike the one just described was noted on May 6, 1949, when a stuffed blue-winged teal dummy was put on the resting place of a pair of teal. After five minutes the pair returned to their area. The drake swam toward the dummy, examined it, bobbed his head a few times and then settled down to sleep on the ditchbank beside the dummy. I have no explanation for the great variation found in the behavior of these male blue-winged teal toward stuffed dummies put on their territories. Perhaps the physiological status of the pair is an important factor in determining the degree of aggressiveness found among the males. In these cases we did not know the breeding history of the pairs well enough to know whether their aggressiveness was correlated with physiological condition.

SUMMARY

1. Home range is the area within which a bird spends its period of isolation between the break-up of spring gregariousness following spring arrival and the reformation of fall gregariousness.

2. Shoveller hen #47-604004 was known to return to the same home range of approximately 200 acres for four consecutive seasons where she produced 65 eggs in seven nests.

3. Attachment to home range varied with sex, and with individuals, but existed in hens longer than in drakes. Some hens were known to remain within the same home range for at least five months.

4. "Territory" has been defined by others as a "defended area." Most territorial study to date has been concerned with passerine birds. Hochbaum's discussion of territory gives three main points: (1) "At the time the pair is ready to nest it takes title to a small water area of the breeding marsh—a pothole, the corner of a slough, or a portion of bay edge. Day after day, as long as the drake and hen remain together as a pair, they may be found on this water area." (2) "The water area occupied by a pair of nesting ducks is defended by the drake; he establishes definite boundaries against the intrusion of other sexually active birds of his own species." (3) "A territory is a specialized piece of terrain in which four components must exist together: Water, loafing spot, nesting cover (adjacent or nearby), and food." Hochbaum makes qualifications to take care of movement outside of the defended area.

5. At Delta, I found that the above concepts of territory could not always be applied to the birds under my intensive observation.

6. Deviations from the general pattern of territorial behavior included the following: Pairs did not always establish definite boundaries to defended areas; drakes defended from several areas, and wide variation between individuals was noted; the various components required in a home range did not always exist inside a defended area.

7. In making counts of breeding pairs as a census method, one encounters wide variations in populations depending on the movement of the birds. It was found at Delta that population figures of one area varied from dawn to dusk, dropping to their lowest number in the evening.

8. A trading of loafing places occurred between pairs. Several

distinct pairs were known to alternate their use of a single loafing place.

9. There appeared to be a close correlation between aggressive behavior and the beginning of nesting.

10. "Territory" sometimes is used to identify an area occupied by a breeding pair without reference to "defense." Such undefended areas were observed at Delta.

11. Understanding the variation in territoriality caused by population density, topography, and the interspersion of land-water types will require the further marking of large numbers of drakes and hens during the breeding season.

12. When we studied reactions of breeding drakes to stuffed dummies in nuptial plumage, a wide variation of behavior was found to exist.

CHAPTER V

NESTING TERRAIN

*W*HEN the first ducks arrive on the breeding grounds, only small patches of open water are available to them. Here and there small snow banks cover frozen ground. Strong, cold winds still sweep the northern prairies and each night much of the open water turns to ice. In many ways the nesting terrain is an inhospitable place for the first returning birds.

The nesting cover used by the earliest-nesting birds consists of dead vegetation from the previous year, as no green vegetation is available until at least a month after the first ducks arrive. Within the nesting terrain there are two main components: (1) the nesting meadows and (2) the loafing and feeding waters which are near.

Nesting cover types.—In the ever-changing complex of plants that make up the nesting meadows, certain key species are important. These vary with locality, soil, and weather. In the region of the Delta marsh, of which my special study area was but a part, the nesting cover plants for 683 mallard, pintail, gadwall, shoveller, and blue-winged teal nests were determined and are given in Table 11.

Certain plant preferences on the part of the ducks are evident in these data. For example, 278, or 41 per cent of all nests found, were located in whitetop or in cover containing some of this plant. Cordgrass ranked second and contained 118, or 17 per cent of all nests found. Bluegrass ranked third, with 86, or 12 per cent. Of the remainder of the cover types, quackgrass contained less than 5 per cent of the nests. These cover plants will predominate or disappear according to land use, water

TABLE 11. COVER PLANT SPECIES FOR 683[1] DUCK NESTS OF FIVE SPECIES

Cover Species	Percentage of nests in each type				
	Mal-lard	Pin-tail	Gad-wall	Shov-eller	B-W Teal
Size of Sample	143	222	38	65	215
Whitetop	58	51	58	28	24
Cordgrass	12	8	18	26	27
Bluegrass	5	10	8	18	19
Quackgrass	0	2	0	20	17
Annual weeds	5	6	15	3	2
Phragmites	11	3	0	0	1
Skunk grass	1	4	0	1	4
Grain stubble	1	6	0	0	0
Brome grass	0	0	0	0	5
Sedges	0	4	0	4	0
Goldenrod-Aster	3	2	0	0	1
Bulrush	4	1	1	0	0
Snowberry	0	1	0	0	0
Willow	0	1	0	0	0
On muskrat house	0	1	0	0	0

[1] This figure includes 90 nests observed by me in a 1939-1940 study at Delta.

levels and other influences. Principal land-use types of the nesting grounds in the Delta marsh region were as follows:

Grain stubble and fallow.—The Portage plains, south of the Delta marsh, were typical of many regions of the Prairie Provinces. It was "grain country"; wheat, barley, flax and oats were the dominant crops. Many mallards and pintails began nesting in spring before the crops were sown, and the occurrence of nests of these species in old stubble was frequent. The amount of cover on the stubble fields was slight (Figure 12). It was not uncommon for pintails to nest in areas so bare that the hen could be seen on her nest from 40 yards away. Pintail nests in such light cover frequently were placed in depressions (Figure 13). Many of the early nests were plowed under at seeding time.

Because of the serious and heavy infestation of the land by sow thistle, wild mustard, and Canada thistle, it was a

Figure 12. The amount of nesting cover on old stubble fields was slight.

Figure 13. On bare fields pintails often placed their nests in depressions.

Figure 14. Phragmites formed dense "jungles" seldom used by nesting ducks.

Figure 15. Only pintails tolerated extremely grazed bluegrass, and often situated their nests beside a stalk of some unpalatable plant left by the cattle.

Figure 16. Even under heavy grazing, unpalatable cordgrass furnished excellent nesting cover.

Figure 17. Heavily grazed sloughs offered attractive edges for loafing ducks.

Figure 18. Ungrazed sloughs with dense edges were not attractive to loafing ducks.

common practice to leave some fields fallow about one year in three. These were plowed throughout the summer to kill the weeds, but no crop was sown. There was almost no vegetation to conceal the birds on these bare stretches. Nevertheless, such fallow fields were used occasionally by nesting mallards and pintails. Rarely plowed early in spring, this land was worked when the farmers had seeded their other acres. Hence, late spring tilling of the land occurred after nesting birds were established, causing nest losses.

On the study area at Delta, where stubble and fallow fields made up 7 per cent of the area, 14, or 6 per cent of pintail nests, and one of 109 mallard nests were found there over a 5-year period.

Grazed pastures.—Comparatively little grassland in southern Manitoba was left ungrazed. It seemed to be the general rule to graze pastures heavily, with the number of cattle probably beyond the ideal. On the study area, 14 per cent of the land area consisted of heavily grazed pasture.

The 800 acres of grazed pasture on the study area was used by about 100 cattle each season. With large blocks of land unsuitable for grazing within this 800-acre pasture, the grazing density was considerably heavier than one cow per eight acres. Evidence that this grazing pressure was excessive was found in the sparse appearance of the grass. During the first four years of this study, bluegrass seldom reached a height of one inch before it was eaten off.

Ungrazed meadows.—Ungrazed marsh meadows formed the most important and productive nesting cover at Delta. It made up 14 per cent of the study area and accommodated 517, or 87 per cent, of the 593 nests found on the study area. The most important cover plant in these meadows was whitetop. Second in importance was cordgrass. Other plants were annual weeds, goldenrod and aster, and quackgrass.

Some of the ungrazed meadows on the area were cut for hay in July and August, but they were grown up enough by fall so that good nesting cover was available again the following spring.

Roadsides.—The narrow strips of roadside that bordered marsh roads were built up above marsh level and therefore were comparatively dry. As a result, a group of plants grew there that were not to be found in the marsh. At Delta, these plants were bluegrass, quackgrass and brome grass. Of the entire 683 nests found, 65, or 10 per cent, were located along roadsides.

Phragmites "jungles."—Heavy stands of tall yellow cane covered large areas in the Delta marsh. On the study area, 41 per cent of the land was covered by this heavy vegetation. In this type, 17, or 11 per cent of all mallard nests were found; 7, or 3 per cent of pintail nests; and two, or less than 1 per cent of all blue-winged teal nests. It contained none of the gadwall or shoveller nests. In all instances, these nests were either in an isolated clump of phragmites, within three yards of some other cover type, or were just at the edge of a large stand of phragmites. I doubt that any of the surface-feeding ducks nest deep within phragmites jungles. The plants sometimes are taller than 8 feet and almost impenetrable by birds or man (Figure 14). The vegetation is so dense that it makes entirely unsatisfactory nesting cover.

Brush and trees.—Along the edges of most Canadian marshes there are some low shrubs and trees. At Delta, the principal trees and shrubs are ash, willow, bur oak, box-elder, cottonwood, dogwood and red-berried elder. I have found these areas to be used rarely by the puddle ducks for nesting. Of the 683 nests, only three (all pintails) were found in this cover type.

Influences on nesting cover, water levels.—The most important factor affecting nesting cover is water level. In 1946, when this study was begun, levels were low in the Delta marsh. They continued to be low in 1947, but in 1948, they responded to a sudden rise in Lake Manitoba. Both marsh and surrounding meadows were affected. Water continued to rise in 1949 and reached a still higher level in 1950. The effect upon nesting-cover plants was apparent throughout this period.

Plants making up the nesting cover at Delta reacted in different ways to rising water. The reactions of some of the most important of these plants to changing water conditions is summarized in Table 12. The line of separation between moisture   conditions is, of course, not a sharp one, and the degree of drought or flooding that plant species will tolerate may vary. For the purposes of this table, three moisture conditions are used: (1) "dry years" when the ground is no longer moist, and heavy vehicles can be driven over it without sinking in; (2) "wet years" when the ground is muddy, vehicles sink into the earth, and a shallow sheet of water covers the ground in spring; and (3) "under flooding" when one foot to several feet of water covers the ground during the spring months.

In 1946 and 1947, no blooming whitetop grass was evident. Quackgrass patches thrived and bloomed on the meadows. Cordgrass did not bloom.

Following the rise in water levels in 1948, large blocks of whitetop bloomed profusely for the first time during the study and continued to flourish in 1949 and 1950. At the same time that the whitetop flourished, quackgrass began to die out. During the wettest year of the study, 1950, cordgrass bloomed profusely for the first time.

Continued high water encouraged hard-stem bulrush, soft-stem bulrush and cattail to take over some areas where water was deep. Thus, in five years, a change from dry non-blooming whitetop and flourishing quackgrass to bulrush and cattail occurred.

Temporary waters function in the same way as plowing or extremely heavy grazing in that they revert plant succession to its first stage. Areas that were submerged temporarily in spring produced wild barley following the June dry-up.

Influence of grazing.—Heavy grazing of a nesting meadow affects the cover in so far as the cover species are palatable to the cattle. During the years 1946 through 1949, bluegrass in the grazed parts of the Delta study area remained short. Only pintails used such light cover for nesting (Figure 15). A pasture of bluegrass becomes almost useless to ducks when grazed heavily. In the other portion of the grazed pasture, however, despite the heavy grazing, plant species unpalatable to cattle thrived and were as good for nesting cover as in ungrazed meadows. Cordgrass (Figure 16), wild barley and Canada thistle made good nesting cover in heavily grazed pastures where the palatable bluegrass was eaten to the ground.

In marshes where such species as phragmites, bulrush and

TABLE 12. REACTIONS OF IMPORTANT MARSH PLANTS TO VARIOUS WATER CONDITIONS IN THE DELTA MARSH, MANITOBA

| Plant | Moisture Conditions and Plant Reactions | | |
	Dry Years	Wet Years	Under Flooding
Phragmites	Survives	Thrives and spreads	Dies out
Cattail	Dies out	Thrives, invades new areas	Dies out
Hard-stem bulrush	Dies out	Thrives	Survives
Whitetop	Survives but does not bloom	Thrives, blooms profusely if ground covered with shallow water in spring	Dies out in some places where water is deepest
Quackgrass	Thrives	Dies out	Dies out
Wild barley	Thrives	Thrives and spreads	Dies out
Cordgrass	Survives but does not bloom	Blooms profusely	Dies out
Willow	Invades marshes	Survives	Dies out
Cottonwood	Invades marshes	Survives	Dies out

cattail choke the edges of water areas, the trampling of edges by cattle destroys the vegetation and creates improved loafing areas for ducks. Where loafing edges were not thus provided in the Delta marsh there was little opportunity for ducks to use many nesting areas because of the absence of this necessary component of the habitat.

Where shorelines are grassy and not overgrown, there seems to be little need for the cattle to make these resting edges. In a study of the drier short grass prairie country of South Dakota, Bue, Blankenship and Marshall (1952:413) concluded that grazing densities should not exceed one cow per 27 acres. They also found that grass-type shorelines supported two to three times as many pairs of breeding water-fowl as did the mud-type bank.

Still another way in which cattle grazing may affect nest-ing ducks has been discussed by Bennett (1937:396). He con-cluded that grazing one cow per six acres in Iowa in normal years appeared to be beneficial to duck nesting areas because he believed that light grazing destroyed some of the skunk and badger habitat and thus eliminated those potential nest predators.

Influence of wild mammals.—Two species of ground squirrels inhabited the nesting meadows and their borders at Delta: the Franklin ground squirrel and the Richardson ground squirrel. The first of these inhabited the brushy marsh borders and played an important part as a predator upon duck nests (Sowls, 1948:130), but exerted practically no in-fluence upon nesting cover. The Richardson ground squirrel inhabited heavily grazed pastures but, so far as is known, seldom disturbed nesting ducks. Its effect upon nesting cover, although of little significance, was great enough to be noticed. Occasionally, areas were made bare of vegetation where this mammal was abundant.

The abundance of the Richardson ground squirrel in-creased with grazing, and this relationship was conspicuous on the Delta study area. No Richardson ground squirrels were seen in ungrazed meadows. The only existing population in the entire marsh, with the exception of the southern marsh border, was in an isolated 5-acre pasture in the village of Delta. The population, completely cut off from any other, maintained itself and flourished where cattle kept the grasses short.

Besides ground squirrels other abundant rodents in the nesting meadows were the pocket gopher, meadow mouse, and woodchuck. With the exception of large mounds of earth which were made by the pocket gopher, no real effect on nesting cover was produced by these rodents. Gopher mounds occasionally covered nesting vegetation over small areas.

Throughout the course of this study, the meadows were used by white-tailed deer, mink, long-tailed weasels, and skunks with no appreciable effect on the nesting cover.

Water area types.—When the ducks arrived in spring, residents and transients alike settled down on the shallow temporary ponds that were scattered over the broad Manitoba prairies. As migration ceased, the density of loafing birds decreased until only the residents were left. These continued to rest on the dry edges, and feed on the abundance of weed seeds and waste grain, well into the nesting season.

Along old creek beds and channels, long narrow sloughs were numerous. These were permanent water areas, and they served as loafing and feeding areas for waterfowl during the nesting season, after the temporary ponds had disappeared. Here the vegetation consisted of a larger and more permanent variety of plants: cattail, bulrush, pondweeds, water milfoil, bladderwort, duck weed, and a variety of other aquatic plants. The edges of the sloughs were either good or poor for loafing birds, depending largely upon the extent to which the area was grazed. Ungrazed edges were dense with vegetation and were, therefore, unattractive to ducks. On the study area at Delta, three such sloughs existed. Only one of these was used extensively by breeding birds. The one used had a heavily grazed edge (Figure 17) as contrasted with the other two which had shorelines heavily grown over with bulrush and cattail (Figure 18).

Gravel roads across marsh areas in southern Manitoba had been made by piling earth and gravel along section lines. The earth had been removed from the sides of the road as the construction crew advanced. When the road was finished, one or two roadside ditches, often full of water, ran beside the road, sometimes for miles at a stretch. The ditches, when located on the prairie, often dried up in summer but in marsh areas remained full throughout the season. The ditches were important to nesting ducks as loafing and feeding places. The roadside ditch on the Delta study area was two miles long and used heavily by ducks during the nesting season. Obviously, not all roadside ditches were of equal importance. Travel disturbance, bare clay bottoms, or dryness often made them useless to the birds.

Since surface-feeding ducks are birds of the open country during the nesting season, they use the large bays and marshes less in spring than in fall. On the other hand, ditches, pot-holes and sloughs have fewer ducks in late summer and autumn than they do during the breeding season. The move-ment of ducks from small water areas to larger water areas follows the end of the breeding season.

Distances from nests to water.—Distances from nests to water were measured for 611 nests of five species during this study. These distances are shown graphically in Figure 19. Although the sample size varies between species, it can be seen clearly that mallards, gadwalls and pintails nest much farther from water than do shovellers and blue-winged teal.

Land-water pattern.—At the beginning of this chapter, nesting meadows and loafing and feeding waters were given as the components of nesting terrain. The usefulness of a breeding area depends upon the amount and distribution of these components, upon their proportions and pattern. Both good and poor land-water patterns were apparent at Delta when the special study area was compared with other parts of the Delta marsh.

Egg-hunting for hatchery purposes had been carried out in most parts of the 30,000-acre Delta marsh. Simultaneously, during the five years of this study, intensive nest hunting was done on the study area. It has been apparent from these studies that the special area, selected mainly because of con-venience, proved to be a better and more consistent nesting unit than any other similar-sized area in the marsh. This was true for the species named: mallard, pintail, gadwall, shovel-ler and blue-winged teal.

One of the reasons for the area's being better than all others of comparable size was its interspersion of good nesting cover and water. The concept of interspersion of cover types for breeding waterfowl follows that for other game. Leopold (1933:451) gives the following definition: "Interspersion. The degree to which environmental types are intermingled or interspersed on a game range." In the Delta marsh, other large

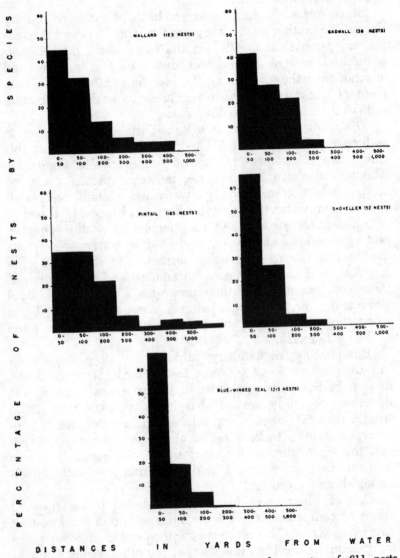

Figure 19. Percentage distribution of distances from water of 611 nests of five species of puddle ducks.

nesting meadows outside the study tract appeared to fulfill all other requirements but lacked adjacent water. A clue to the relationship between water and nesting area can be found when we consider the distances from water that the nests of various species were located. Distances from water of 611 puddle duck nests have been given by species in Figure 19.

Difference in quality of areas.—On the Delta study area, where the long roadside ditch is flanked on both sides by nesting cover, and beyond this by sloughs and bays, certain sections were consistently better waterfowl producers than other sections of equal size. To demonstrate this difference, I selected a section of the roadside ditch at the southern end, which I henceforth shall call area "A" and a section of the

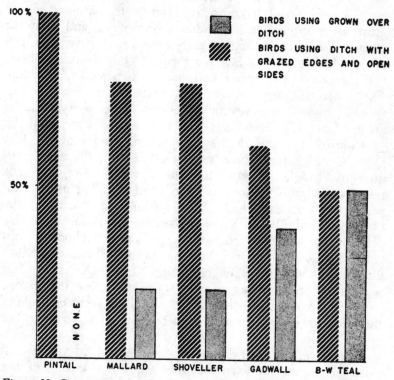

Figure 20. Comparative populations of five species of puddle ducks on two equal lengths of ditch during 1947 based on 22 counts.

same length at the northern end, area "B." Whereas A was characterized by good loafing areas, with large adjacent nesting meadows and grazed edges, B had little or no loafing areas but had adjacent heavy cover of phragmites and grown-over ditch banks.

Figure 20 shows comparative indices and includes a base figure of the total number of ducks of each species counted from March 23 to July 20, 1947, on these two areas. The dry year of 1947 was selected for this comparison to avoid the complicating influence of occasional flooding which occurred in area A when ditch banks overflowed. Only counts made of the entire ditch without intervening delays were used.

When populations of resident birds were compared, it was clear that pintails preferred area A and, during 1947, did not use area B at all. Being almost exclusively a bird of the open, pintails will not tolerate heavy stands of phragmites and grown-over ditch banks. Mallards, shovellers, and gadwalls preferred area A, too, but tolerated the heavy cover and lack of good loafing areas of B to some extent. Only the blue-winged teal showed no preference between the two sections of ditch. The short distance between nest and water areas of this species made large nesting meadows unnecessary, smaller areas of nesting cover, such as roadsides, being adequate.

Control of undesirable vegetation.—Undesirable plant growths which form monotypic vegetational patterns are unattractive to wildlife. At Delta, yellow cane or phragmites has taken over large sections of marshland, forming almost impenetrable jungles, which ducks seldom use except at the borders. Ward (1942) has discussed the management of this plant in the Delta marsh and suggests two methods of control: (1) mowing and (2) burning. Both methods are of limited value. Mowing is effective only if repeated year after year until other plants succeed in taking over the marsh. Ward has pointed out that fire is not effective in controlling this plant where the ground is wet but is effective when the ground is dry.

Large gaps in our knowledge of marshland plants still exist. We have, for example, only vague knowledge of the plant succession patterns of marshland meadows. Without the basic knowledge of marsh plant requirements, and their reactions to environmental changes, we cannot control them intelligently or change them to increase the duck carrying capacity of a marsh.

Agriculture and waterfowl management.—The marsh edges at Delta, where agricultural land and wild marsh meet, are extremely important as nesting areas of the surface-feeding ducks. This is true of most marshes in the Canadian prairie provinces as well as of areas in the northern parts of the United States.

Agricultural practices, as they affect these waterfowl breeding areas, are many. In dry years, farmers invade the marsh edges to grow flax and other small-grain crops. In wet years, they retreat from these areas and let wild marsh plants take over. Ordinarily, however, the farmers cut wild hay in scattered areas of the marsh that are dry enough to support horses and haying equipment. Cattle usually are grazing these wet marshlands.

As the farming ventures see-saw back and forth between wet years and dry years, so does the area used by waterfowl. In wet years, pintails and mallards nest far out on the prairies, but in dry years, the prairies are devoid of breeding ducks. On these marsh edges the weather dictates to farmer and to waterfowl which areas each will have. Aside from this, however, there is much that the farmer could do to manipulate the land to favor ducks without neglecting his own economic interests.

One of the first things that the farmer could do is to arrange his marsh burning so that duck nests are not destroyed and so that much desirable duck nesting cover is not burned off.

Where land is dry enough for consistent hay cutting, no problem exists. But because of the changing water levels, many areas are cut only occasionally, and in these areas a build-up of dead materials occurs. The heavy mat of dead vegetation from a former year's growth makes the cutting of hay almost impossible, and to combat this situation and make sure that the summer's hay will be all new growth, many farmers burn the areas in spring. A farmer could aid waterfowl production in this regard by burning early in the spring before nests are abundant. Also he could aid waterfowl production by burning selectively only the areas where he intends to cut hay later, rather than burning large meadows indiscriminately.

Fire lanes designed to augment natural barriers to fire, are desirable and necessary in our marshes. Maintenance of fire lanes and supervision of the burning of marshlands should be a legitimate expenditure of game management funds. Considerable headway in control and education as applied to marsh burning has been made by Ducks Unlimited (Canada). Further progress on an extensive scale is needed.

The farmer and rancher can benefit waterfowl production further by conservative grazing of breeding marshes. It has been shown already that cattle grazing is an effective management tool in the creation of loafing edge for waterfowl.

Excessive numbers of cattle will damage grasslands for the cattle, however, at the same time that they damage duck breeding areas. As a general rule, conservative stocking of cattle for the good of the range will benefit ducks.

SUMMARY

1. The components of nesting terrain included (1) the nesting meadows and (2) the loafing and feeding waters which were near.

2. Data on cover types for 683 nests of mallard, pintail, gadwall, shoveller and blue-winged teal showed that, at Delta, whitetop was the most important nesting cover plant.

3. Mallards and pintails often nested on the bare grain stubble fields and sometimes on bare fallow ground. Many of these nests were destroyed by farmers during early spring plowing.

4. Most pastures in Manitoba were so heavily grazed that little, if any, nesting cover was left for ducks except where plants unpalatable to cattle occurred (wild barley and cordgrass).

5. Heavy phragmites formed dense jungles which rarely were used by nesting ducks.

6. Water level was the most important factor influencing the vegetation of nesting meadows. In dry years, cattail and bulrush died out; trees invaded the nesting meadows; phragmites, quackgrass and wild barley thrived; and whitetop survived but did not bloom.

7. In wet years, short of flooding, cattail and bulrush thrived;

trees survived; phragmites thrived and spread; quackgrass died out; wild barley thrived and spread; and whitetop and cordgrass prospered and bloomed profusely.

8. When meadows remained flooded, quackgrass, cordgrass, trees and phragmites died out. Cattail and whitetop survived in places, and only hard-stem bulrush survived under heavy flooding.

9. Wild mammals had little effect upon the nesting cover of waterfowl at Delta.

10. Water areas used by ducks during the breeding season included potholes, sloughs, ditches and large bays. Large bays were used less by surface-feeding ducks during the breeding season than they were later in the summer and autumn.

11. The value of a marsh for breeding waterfowl depends upon the amount, quality, and distribution of the required nesting, loafing and feeding areas.

12. Information obtained on the distances of 611 nests of five species from water showed that the mallards, pintails, and gadwalls nested over a wider area, and farther from water, than did the shovellers and blue-winged teal.

13. Nesting cover more than 100 yards from water accommodated only 10 per cent of the shoveller nests, but had 24 per cent of the mallard nests, 31 per cent of the pintail nests, and 29 per cent of the gadwall nests.

14. Ungrazed ditches that had heavy vegetation on their banks and a shortage of dry nesting meadows adjacent to them were used less by ducks than those with bare, grazed edges. Pintails showed the greatest preference for open bare banks; mallards were second; shovellers third; gadwalls fourth. Blue-winged teal used the areas with grown-over ditch banks as frequently as they used those with grazed edges and an abundance of adjacent nesting cover.

15. For blue-winged teal, the narrow strip of roadside grass was adequate as a nesting place.

CHAPTER VI

NESTING SEASON

*E*ACH YEAR the waterfowl species have the opportunity of regeneration. The time of year for this regeneration is known as the nesting season. It is upon this period primarily that the survival of each species depends; it is also upon this period that waterfowl management depends for its annual crop of harvestable birds. In the annual cycle of events, populations reach their yearly low point just prior to this time; their annual peak in numbers is just at its close. It is important, therefore, to understand this period as completely as possible.

While most of this study deals with some phase of the nesting season, this chapter is devoted to *time* alone. Thus here, I will discuss the time of year that breeding occurs, the length of the nesting season, how it changes from one year to another, and how it varies between the different species of ducks.

The over-all span of the nesting period at Delta has been discussed by Hochbaum (1944:94) who illustrated graphically the laying dates of the species that nest there.

In this study an attempt was made to determine the length of the nesting season in order to show the differences beeen years as well as the variation between species. An attempt also was made to determine the effects of weather upon the time of beginning and length of the nesting season.

A choice of methods was open to me. For example, masses of brood data might have been gathered and the ages of broods plotted to show the results of nesting activity. Had this been done, we would have obtained an erroneous picture of the occurence of nesting peaks. Obviously that method would have taken no account of predation on early nests, and the picture would have shown only late successful nesting activity. Indeed, 90 per cent of all first clutches might have been laid and destroyed without altering the picture at all; and only the 10 per cent successful nests which brought off broods would have been indicated in our graphic picture.

In order to create a clearer and more accurate picture of the nesting season, I used the nest rather than the brood as a starting point. The routine was to find nests, determine their beginning dates, and plot these nest-startings by weeks and by species.

In order to determine the starting date of each nest, it was necessary to age the embryos and back-date to the day the first egg was laid. The aging of embryos involved the breaking of one egg from each nest and comparing it with a collection of known-age embryos which I had prepared in the hatchery. Since a hen normally laid one egg each day, the formula was: number of eggs in nest, plus number of days of incubation, equals the number of days since the first egg was laid. For example, a nest of 10 eggs was found on May 20. It was determined to be in the tenth day of incubation. Hence, 10 (number of eggs) plus 10 (number of days incubated) equalled 20 (number of days since first egg was laid). Thus we knew that this nest was started on May first.

By this method, mistakes of a few days may have occurred but, despite these, it proved to be a far better method than any other for determining the starting dates of nests. Errors probably occurred because of several reasons. Hens may have dropped eggs outside the nest when disturbed at laying time. When this happened our calculations must have been thrown off one day for each egg dropped. It was not always possible to determine whether a laying hen had deposited her egg for

the day when her nest was found; hence, a one-day error could have resulted in this manner. Finally, the aging of embryos for field work was not accurate to within less than a day.

These errors are of small importance in comparing the nesting chronologies of several years, and are not important in comparing the nesting chronology between species for any particular year.

This technique was used in 1949 and 1950. Figure 21 shows the nesting chronology of the mallard, pintail, and blue-winged teal for the years 1949 and 1950, as determined by the percentage of nests started by weeks of the season.

Figure 22 shows the chronology of the shoveller and gadwall nesting seasons based on the number of nests started during the different weeks of the season.

Description of data.—For the years 1949 and 1950, starting dates for 64 mallard, 15 gadwall, 98 pintail, 29 shoveller and 112 blue-winged teal nests were determined. The volume of data for the mallards, pintails and blue-winged teal is sufficient to give a good curve of the nesting chronology for these years. The small quantity of data for the shovellers and gadwalls allows limited interpretation.

Comparison of nesting dates.—From these figures we see that the mallard and pintail nesting peak was reached before the last three species began nesting. This was the usual situation in southern Manitoba. The mallards and pintails were the early-nesting species, and the gadwalls, shovellers and blue-winged teal always were late-nesters. Just as the pintails and mallards were first to begin nests (one or two weeks before the others), they likewise were first to terminate nesting.

Comparison of nesting seasons.—In the opening chapter, I discussed the arrival dates of the various species and pointed out that they varied considerably from one year to the next. The variation span in arrival of the mallards and pintails was two weeks; of the gadwalls, shovellers and blue-winged teal, one week. Just as the arrival dates varied with years, so did the beginnings of nestings vary. The spring temperatures of 1949 were near normal for southern Manitoba, but the spring of 1950 was one of the latest on record. The difference in nesting chronology between these two years was conspicuous. To illustrate the difference, I have plotted nesting data by species.

It is apparent from Figures 21 and 22, that the mallards and pintails began nesting two weeks later in 1950 than they did in 1949, and extended their nesting season by about the same length at the end. The blue-winged teal began nesting one week later in 1950 than in 1949 but ended the nesting season at about the same time both years. The shovellers began nesting during the same week in 1949 as they did in 1950. The peak of the nesting season, however, was reached two weeks later in 1950 than in 1949. There was no perceptible difference in the close of the season.

Hence we see that of all species studied, the shoveller was the only one which started some nests during the same week each year. Consistent with this evidence is the 4-year record of an individual shoveller hen given in Table 13. She began nesting in the late spring of 1950 even earlier than in the early spring of 1949.

Small numbers of gadwall nests were found during the last week in April in 1949, but the majority were found that year the first week in June. This seemed to indicate that the nesting of this species was delayed by the cold weather and that most of the hens began nesting at about the same time.

When we try to interpret these data it is more difficult to understand the end of the nesting season than the beginning. For all species, including the gadwall and shoveller, for which there are only meager data, I know that the beginning date demonstrates, for each year, the date when nesting actually began. For the end of the season, however, the picture is less clear. Only for the mallard, pintail and blue-winged teal are there enough data to warrant a conclusion. For these species it seems likely that an extension of the nesting season existed in 1950 comparable to the delay which occurred at its start.

Temperatures and the nesting season.—Just as it has been shown that temperature is an important factor in determining the arrival dates of waterfowl, it also can be shown that temperature is important in determining the beginning of nesting. It is clear that nesting began in 1949 much earlier than in 1950. A look at the weather record for the period shows us that the 1950 delay in nesting coincided with continued low temperatures.

A comparison of mean daily temperatures for these two years for the period April 15 to May 30, and the beginning nesting dates of five species of ducks for these two years, are shown in Figure 23.

The mean daily temperature of a "normal" year for this period was 47 degrees Fahrenheit. In 1949, the mean daily average was 50 degrees Fahrenheit, and in 1950, it was only 43 degrees Fahrenheit. Furthermore, 1949 had no prolonged below-normal periods; whereas, 1950 had well-below-normal temperatures most of the time.

The period April 15 to May 30 is an arbitrary one selected because most first nestings probably begin within this period.

The effects of temperature on the beginning of nesting are seen when we compare the beginning dates for the two years by species.

TABLE 13. FOUR-YEAR NESTING RECORD FOR SHOVELLER HEN NUMBER 47-604004

Year	Date 1st nest started	Number eggs 1st clutch	Number eggs 1st renest	Number eggs 2nd renest
1947	May 22	12	8	8
1948	May 21	12	?	?
1949	May 18	1	12[1]	?
1950	May 16	12	?	?

[1] A continuous laying record.

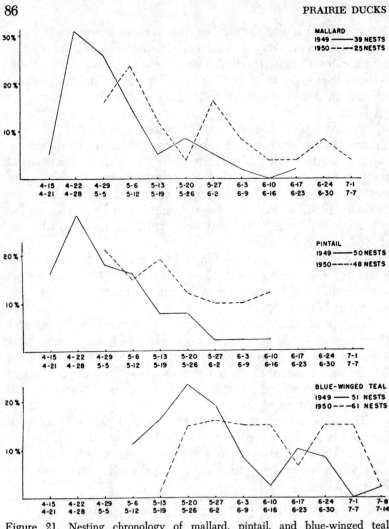

Figure 21. Nesting chronology of mallard, pintail, and blue-winged teal for the years 1949 and 1950 determined by the percentage of nests started by weeks of the season.

The effect of temperature on all activity during the spring was observed easily both in the wild and in the captive birds on the station pond. On warm evenings, following the first arrival of the birds, we saw mallard and pintail pairs cruising over the marsh, apparently selecting nesting locations. But on cold evenings, this activity ceased. In the hatchery pond, where we maintained a flock of hens, there was constant

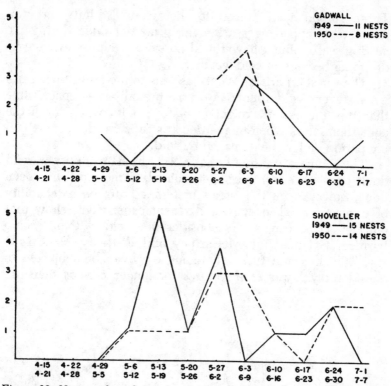

Figure 22. Nesting chronology of gadwall and shoveller for the years 1949 and 1950 determined by number of nests started by weeks of the season.

courtship activity on the part of the wild drakes which came into the pond. This activity reached its peak on warm evenings, but was retarded greatly or ceased when the temperature dropped to near freezing. Apparently there is a temperature threshold just above freezing where activity stops. At exactly what temperature the threshold exists is difficult to determine. Although we have not been able to measure this precisely, we have been able to predict changes in activity because of temperature changes. A mild evening following a cold period brought on the resumption or beginning of this activity. Temperature fluctuation from one day to the next, and during one day, coincided with considerable activity or the complete absence of it.

I heard rumors that ducks were dropping eggs on the edges of prairie ponds during cold weather without building nests.

These rumors were traced to their sources, but were unfounded. Dropped eggs are a fairly normal thing during the nesting season, but I have found no evidence of excessive egg-dropping because of cold weather.

The question arises: What happens to hens and their nests when severe cold follows a warm period? It is conceivable that the temperature might drop to a point where even the eggs themselves when unguarded (as they are during the laying period) might be damaged by cold.

The temperature limits that affect the hatchability of wild duck eggs have not been determined. Some work has been done, however, on the effects of temperature on hatchability of domestic chicken eggs, and recent experiments show that chicken eggs can stand considerable cold without being harmed. Of this, Jull, McCartney and El-Ibiary (1948:140) say, "Chicken and turkey hatching eggs can be subjected to considerably lower temperatures for longer periods than has

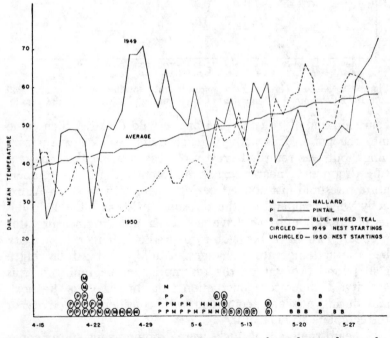

Figure 23. Relationship of temperature to start of egg-laying during the years 1949 and 1950 for mallard, pintail, and blue-winged teal.

usually been considered possible without seriously affecting hatchability."

During a cold spring at Delta, it was not uncommon to hear tales of great losses of duck eggs from freezing. I have found no evidence, during the study, of frost causing a loss of wild duck eggs. In the spring of 1950, however, a temperature of 20 degrees Fahrenheit on April 26 caused some loss of Canada goose eggs in the nests of a captive flock kept at the station pond. These nests were not being incubated, and, when exposed to this temperature, some were frozen and the shells cracked. During this late spring no mallard nests were known to have been started before May 1, and no pintail nests before May 2. Hence no wild duck nests were available for observation during this period.

The goose nests here discussed were exposed with little or no vegetation surrounding them. The extent to which ground cover and nest lining protects duck eggs from freezing when the hen is not on the nest probably is great. The temperatures at which duck eggs freeze, and the importance of low temperatures during abnormally late springs in determining production, should be studied more thoroughly.

The most significant effect of low temperatures seems to be the delay in the beginning of the nesting season.

End of the nesting season.—Nest hunting during each year of this study was begun as soon as the first ducks arrived in spring and continued until after repeated searches yielded no nests. In addition to nest data, there were other means of determining the end of the nesting season, and it also was possible to get information about the percentage of hens which were unsuccessful nesters. In 1950, the end of the nesting season was followed closely by an influx of pintails into the Delta marsh. This population did not consist entirely of males as would have been the case had all hens still been renesting. For one small water area within the study area, counts of pintails were begun as soon as the influx became apparent and were continued until the drakes were so far into eclipse plumage that they could no longer be distinguished from females. These hens apparently were unsuccessful nesters which were no longer trying to renest, as all successful hens should have been with broods or on nests.

SUMMARY

1. The beginning dates for nesting were determined by taking an egg from each nest found, by aging the embryo, counting the number of eggs in the nest, and back-dating to the time the nest was begun.

2. During 1949 and 1950, beginning dates of 64 mallard nests, 98 pintail nests, 15 gadwall nests, 29 shoveller nests and 112 blue-winged teal nests were obtained.

3. These data showed that in 1949 the mallards and pintails began nesting during the week of April 15-21, but in the cold late spring of 1950, they did not begin until about two weeks later.

4. There is some indication that in these two species the season was extended at the end by approximately the amount it was delayed at the beginning.

5. For the blue-winged teal, a later-nesting species, the delay was only one week. The nesting season of this bird apparently was extended by the amount of time it was delayed.

6. The shoveller showed no appreciable difference in nesting dates between the two years during which data were collected.

7. Gadwalls apparently were delayed by at least two weeks in 1950, as compared with 1949. There was no perceptible difference in the close of the season for this species during the two years.

8. The lateness of the season in 1950 did not appear to reduce nesting opportunity seriously. For the early and comparatively rapid-developing species, the season was not so late that large numbers of young would be unable to fly at freeze-up time.

9. The most significant effect of low spring temperatures seemed to be the delay in the beginning of the nesting season. No evidence was obtained in this study to indicate that large losses of eggs were caused by freezing.

10. Large numbers of pintail hens were seen in mixed gatherings as the nesting season came to a close. These hens undoubtedly were unsuccessful nesters which were no longer trying to renest, since all successful hens should have been with broods or on nests.

NESTING BEHAVIOR

TERRITORIALITY, one phase of bird behavior, can be seen in aggressive actions between individuals and the resulting submissive responses that follow. Other phases of bird behavior concern not only the reactions of birds to other birds, but their responses to their inanimate environment. As the displays of one bird incite certain psychological and physiological responses in other birds, so do certain inanimate features of the birds' environment bring about definite responses.

In this chapter, phases of behavior, other than territoriality, will be considered: the selection of the nest site, the building of the nest, the addition of nesting material, the building of the nest canopy, the rate and time of egg laying, the behavior of the drake during the laying and incubating periods, the hen's inability to identify her eggs or to discriminate between addled and fertile eggs, the desertion of the nest by the hen, the courtship behavior of the incubating hen which leads to re-pairing and renesting, and the activity of the laying and incubating hen at her nest.

Behavior of the laying hen.—Evening flights of paired mallards and pintails were common on the Delta marsh at the beginning of the nesting season. These pairs flew low over the marsh, usually at a level of between 10 and 20 feet. The pairs made continuing circles and returns within a limited radius of about one-quarter of a mile. We came to expect these flights on warm evenings in April just prior to finding the first nests of the year. These birds seemed preoccupied and were tolerant of disturbance, flaring only momentarily, then returning to their flights.

The flights appeared to be search-flights, the pairs apparently looking for suitable nesting places, feeding grounds, loafing spots and the other constituents of home range. To many of the old hens, the terrain was familiar because of the experience of former years. To some of the young hens, although less familiar, it was not new.

The question arose as to the interval of time between the evening flights and the first nesting. In order to determine the duration of this interval, I collected six mallard hens from isolated pairs during late April, 1950. These hens were exhibiting the evening flight behavior described above.

In five of the females, the largest follicles were the size of a pea, or about 8 millimeters in diameter. In one, there were four follicles larger than a pea, running up to 22 millimeters in diameter. Although we cannot make close estimates of when these hens would have been able to lay, it was apparent from the anatomical evidence that the evening flight behavior of these hens would have preceded by several days the laying of the first egg.

As soon as egg-laying was in full progress, evening flights ceased. Why such flights do not occur later on, following repairing and prior to renesting, I do not know. Perhaps, by then, the hens are familiar with their area, and the same loafing and feeding places will be used again. As we shall see later, the second nest was located typically in the same general area as the first nest.

The fact that cold weather delayed or brought temporary cessation of the evening flights of pairs, as well as a delay in nesting, indicated a close tie between the birds' physiological status and the weather. The details of this relationship are little known but are worthy of further research.

Building the nest.—Unlike the nest building of most birds, the construction of the nests of surface-feeding ducks does not precede, but rather accompanies egg-laying and incubation. In this detail, the surface-feeders also differ from the diving species. Hochbaum (1944:47) says of the canvasback that the hen often begins building and often abandons several nests before laying, and that nest construction begins two or three days before the first egg is laid. A similar account of the time of laying in relation to nest construction by the redhead is given by Low (1945:50), who says, "Construction of nests often was begun two or three days to a week before egg-laying started."

Evidence at Delta seemed to indicate that the only preliminary preparation by puddle ducks to the laying of the first egg was the construction of the nest bowl or "scrape." For all of these species, it was essentially the same except for size, which varied with the size of the duck. The scrape was a simple, shallow, hollowed-out depression in the ground from which vegetation had been partly removed. The task of making this depression appeared to have been accomplished by scratching (as evidenced by claw marks in the earth). Over the deep scratches, the base of the scrape was worn smooth from the movements of the hen's breast and belly against the moist ground. It was in these bare nest bowls, with only a small amount of nesting material added, that the first eggs were deposited. Even after the clutch had been completed, the first egg laid often retained a coat of mud picked up as it rolled about in the nest bowl under the duck's body.

The time during which the nest bowl is dug is unknown. It may precede the laying of the first egg by several days. Of this, Merrill C. Hammond of the U. S. Fish and Wildlife

Service wrote me (letter dated December 20, 1949): "Data gathered in 1949 at the Lower Souris Refuge in North Dakota suggests that mallards were digging six days before first egg-laying and pintails four days before; . . . on May 13 . . . I suspect that some gadwalls were digging nest bowls. This date was four days prior to first egg-laying for gadwalls."

Addition of nesting material.—The nests of puddle ducks were constructed entirely of material that lay within reach of the hen as she sat on the nest. There was no evidence that the surface-feeding ducks ever carried nesting material from points beyond their reach as they sat. On several occasions, I have taken nesting material from nests deliberately, forcing the hen to reach more material, and found that, as more and more material was used, the density of the surrounding vegetation became thinner. I have watched the extraction of nest material as I sat in a blind and have seen hens reach far out with open bills, grasp leaves and stems, pull them toward the nest, (Figure 24) and tuck them around the nest perimeter (Figure 25).

Building the canopy.—When located in heavy grass, nests of puddle ducks invariably had a well-constructed canopy. The bending of grasses to form this hiding structure obviously was accomplished by the hen as she sat. It appeared that the purpose was to hide the hen and offer shelter from wind, sun, and rain. On July 7, 1949, I planted a clump of grass near a shoveller's nest beside which I had placed a blind. Soon after the hen returned and settled on her nest, she began pulling the grass tops over her (Figure 26). After each pull, the grass was held in place for about five seconds (Figure 27).

Rate of laying.—Laying ducks normally deposit one egg a day. This fact has been accepted generally and was borne out by the data collected in these studies. Of 18 nests representing nine blue-winged teal, six pintails, two mallards and one shoveller, I found that the number of eggs in the final clutch equaled the number of days taken to complete the clutch. There were no exceptions to this rule in any case where a complete history of egg-laying was obtained.

Time of day for egg-laying.—Most mallards. pintails and gadwalls went to their nests to lay between sunrise and four hours after sunrise; most records were obtained during the first half of this period. For these three species, the actual observation of the hen's arrival in the meadow was the best indication of laying time. These were the relatively mobile species which came from distant water and loafing areas to lay and for which pair counts and lone-drake counts were difficult to obtain. Among the less mobile blue-winged teal and shovellers, however, pair counts and counts of lone drakes during the early days of the laying period indicated that the hens of these species normally went to their nests during the latter half of the forenoon. This conclusion was borne out by the fact that early-morning watching to locate nests (as described in Chapter I under techniques) yielded comparatively few nest locations for these species.

Behavior of drake, accompanying hen.—The hen was accompanied by her drake when she was searching for a nest-site and when she was making her daily trip to her nest to deposit an egg. Invariably she led and he followed. At all times during the laying period, the drake accompanied his hen when she was not on the nest. Even when she was sitting on her nest, he frequently waited in the same meadow and sometimes flew over the nest-site in an apparent search for her.

During this period, I often saw lone mallard and pintail drakes circling a particular spot in a nesting meadow and occasionally calling. Following such a flight, the drakes usually returned to an adjacent loafing bar to sleep or wait. I have found nesting hens by watching these drakes and later searching the spot over which they were calling. Thus, the drakes were found to be circling and flying over the spot where each knew his hen to be. Although the calls of the drakes were insistent, the females never were known to answer. The mallard drake's call was the usual low quack, less audible than that of the hen's as I heard them on their evening flights preparatory to nesting. The pintail drake's call was the low double whistle which has been described by C. Oldham as a *quuck quuck* (Witherby, Jourdain, Ticehurst and Tucker 1939:271).

Dissolution of pair status.—That mallard, pintail, and gadwall drakes leave their hens a few days after the clutch is completed was apparent from observations of paired birds, of lone drakes waiting for hens, and of bachelor bands of drakes. The shoveller and blue-winged teal drakes, on the other hand, remained paired far into the incubation period; although they usually left their hens before the eggs hatched.

Behavior of incubating hen.—When eggs are infertile or when the embryos are killed (as occasionally happens when fire sweeps a prairie marsh), hens often sit for long periods beyond the normal hatching date. This is true of many kinds of wild birds. There are numerous written accounts of such behavior. Table 14 gives the number of days some species of birds have been known to sit on dead eggs.

Bird (1930:418) says, "One pintail hen whose eggs were scorched on May 2, continued to sit on them until the first week in August." My longest record was of a pintail which sat on two dummy eggs made of wood from April 28 to June 28, 1949, a total of 62 days, or 41 days beyond the normal incubation period of 21 days. Another pintail hen sat on dead eggs, which had been substituted for her own, from May 17 to July 15, 1950, a total of 60 days, or 39 days more than the normal period. On the sixtieth day, the rotten eggs exploded from heat when the hen was kept off her nest by disturbance. Desertion followed.

Similiar behavior has been found to exist in mallards, gadwalls, shovellers, and blue-winged teal where hens were kept incubating for shorter periods of time beyond their normal incubation periods.

TABLE 14. LENGTH OF TIME BIRDS HAVE BEEN KNOWN TO SIT ON DEAD EGGS

Species	Approximate incubation period in days	Days bird sat	Authority
Barn owl	14	84	East (1930:6)
Eastern bluebird	13-15	21	Laskey (1940:188)
Eastern bluebird	13-15	33	Thomas (1946:156)
Eastern bluebird	13-15	36	Swanson (1926:339)
Carolina chickadee	11	24	Odum (1942:430)
Western crow	16-18	32	Emlen (1942:150)
Bobwhite quail	21-23	56	Stoddard (quoted by Leopold, 1933:367)
Hungarian partridge	24-25	32	Sprake (quoted by Leopold, 1933:367)
American pintail	21	62	Sowls (this study)

From the standpoint of management, the tendency of ducks to sit on dead eggs would have significance only when it occurred on a large scale. Abnormally cold weather and heavy rains during the laying season might addle large numbers of eggs; and, if the eggs were not cracked, hens might continue to sit for long periods. The only record that has come to my attention of several hens sitting for long periods was given me by M. C. Hammond (letter dated December 20, 1949) who said that, following a heavy frost on June 8, 1937, in North Dakota, he found six sets of dead eggs with hens sitting on them past the normal incubation period. One pintail sat for 30 days; another 48 days; a mallard, 49 days; a shoveller, 45 days; and another shoveller, 42 days. A baldpate sat 46 days.

How often hens return to burned-over nests is hard to determine. Leedy (1950:234) gives an account of a mallard and a black duck returning to burned nests where at least part of the eggs had been scorched severely and the embryos probably killed. It was not known how long the hens remained on these nests, because they later were destroyed by predators.

Since we do not know how frequently hens sit on addled eggs, it is difficult to evaluate the effect of this behavior on the production of young; but I conclude that it could be an important factor in areas where weather or fire are particularly destructive of eggs.

Nest desertion.—All but one of 18 mallard nests having four eggs or fewer were deserted by the hens as a result of our disturbance. Only one of 12 nests was deserted when the hens had already laid between five and eight eggs. No desertions from flushing occurred after the eighth egg was laid or during incubation. The pintail showed a lesser tendency to desert, even at the early stages. Ten of 20 pintail nests found prior to the laying of the fifth egg were deserted after the hens were flushed once. Only three of 27 pintail hens deserted after being flushed once between the laying of the fourth egg and ninth egg. No pintails deserted after being flushed from nests during incubation. The blue-winged teal showed only a slight tendency to desert after being flushed from nests during incubation.

Hence we can say that desertion resulting from disturbance is most likely to occur during the early part of the laying period. It occurs most frequently with the mallard and least frequently with the blue-winged teal.

Since these figures represent only the hens which failed to return to their nests after being flushed once, they do not indicate how much disturbance hens will tolerate. It has been suggested by Leopold (1933:366) that a desertion limit may exist, a certain time during the laying period or the incubation period at which the hen would no longer desert after disturbance. It must be remembered, however, that desertion varies with the degree of disturbance and with the individual bird as well as with the species.

Courtship.—Two periods of courtship have been described by Hochbaum (1944-22) for the species of ducks breeding in the Delta region: ". . . the competitive *prenuptial* courtship preceding pairing, and the possessive *nuptial* courtship of the mated pair."

Because it differs from these two, and stems from different causes, a third courtship will be described here and called *renesting courtship.* Only occasional prenuptial courtship of surface feeding ducks was seen at Delta because most of the birds were paired when they arrived and already were in

Figure 24. A shoveller hen reaches for grass to line her nest . .

Figure 25. . . .and tucks the grass into the nest perimeter.

Figure 26. A shoveller hen reaches for a live blade of grass. . .

Figure 27. . . . pulls it over her back and holds it there for several seconds.

Figure 28. A shoveller hen reaches far out to retrieve an egg that has been placed outside her nest. . .

Figure 29. . . . and tucks the egg beneath her by using the under side of her bill.

Figure 30. A shoveller hen flys away from her nest with an eggshell.

Figure 31. A pintail hen hesitates with an eggshell before flying away with it.

their nuptial period. An exception to this was the baldpate, in which prenuptial courtship was common at Delta in late April and May each year. The prenuptial courtship activities of the puddle ducks consisted of various displays, postures, and pursuit flights as described by Millais (1902), Wormold (1910), Lorenz (1941), and Delacour and Mayr (1945). As the last two authors have pointed out, courtship pursuit flights often are confused with pursuit flights associated with territorial behavior. The two flights were distinguishable at Delta. For the first few weeks of the breeding season each year, flights of mallards, pintails and gadwalls were limited to flights associated with territorial behavior. These occurred in mallards and pintails at the time baldpates were making aerial prenuptial courtship flights.

As the season advanced, a third type of flight was seen. This was the renesting courtship. It was evident that the hen initiated the flight by a series of teasing calls. In the pintail, a lone hen uttered a teasing call and threw her head along her side. This undoubtedly was the call which Heinroth (Witherby et al; 1939) termed the incitement note in the European pintail and described as *rarrerrer*. The actions of the hen followed closely those described by Lorenz (1941) in his discussion of the sexual incitement notes and actions of both mallards and pintails. I have watched pintail hens teasing drakes (see frontispiece) in the meadow for a half hour or more at a time. On each occasion, short flights of from 10 to 50 yards, hovering, teasing calls, and head throwing kept several drakes alert and drew them to the hen. Following a series of calls the hen made a towering flight, continuing her calling intermittently. As she did so, she hunched her body, drew her wings in, and dropped momentarily in mid-air as her head went back. I have seen pintail hens climb to what appeared to be over a thousand feet above the prairie on such flights. As the flight began, a number of drakes pursued; the number decreased as many of the drakes dropped out.

Each year at Delta, we observed these late-season aerial courtships of mallards and pintails, although they were absent during the first three weeks of the nesting season. Hochbaum (1944:16) attributed this to the late arrival of unpaired birds and says: "The single drake is a rarity in April as sex tallies of hundreds of birds show an almost perfect 50:50 sex ratio. In early May, however, unmated mallards and pintails reach Delta, such late transients showing a preponderance of males. Hereafter drakes outnumber hens, and one sees courting parties of four or five to a dozen or more drakes in pursuit of a single hen."

Evidence, from my observations of marked individual hens whose nests I have robbed, indicates that this late courtship is associated with re-pairing of hens which have lost their nests after their drakes have left them; and does not necessarily involve late arrivals. In 1948, three marked pintail hens, whose nests I had robbed, were seen "teasing" drakes and were found to be renesting later on the study area. In these three instances, the original drakes had abandoned their hens by the time I robbed the nests. I never have seen courtship flights follow nest destruction when the hen and drake still were paired.

In addition to behavior observations of marked individuals, I have collected hens which were seen teasing drakes, and examined their ovaries for the presence of ovulated follicles. Although the pattern of ovulation in waterfowl is not clear enough to enable us to count precisely the number of eggs that a hen has laid, as done with pheasants by Buss, Meyer and Kabat (1951:32-46), the presence or absence of ovulated follicles can be determined as an indicator of whether or not hens already have laid the same season. Four hens of known behavior were collected in 1948 (three pintails, one mallard). Three of these showed signs that they had laid, but I was unable to tell whether or not the fourth hen had laid that year.

It was noticed that the incitement call of the teasing behavior sometimes could be induced. Upon approaching the nest of an incubating bird whose drake was no longer with her, I came to expect the incitement call as she flew off in alarm and found that this behavior (uttering the incitement call) was, in fact, stimulated by my disturbance. There were no drakes near; and, since the bird was, in each instance, surprised by me, the behavior must have been caused or set off by my approach.

No part of the teasing behavior ever was seen in laying

hens which still were paired with their drakes or in incubating birds which left their nests normally and without disturbance.

There exists in incubating hens (no longer paired) a latent teasing ability which will attract drakes. The similarity of the incitement call and the disturbance call led me to believe that they were, in fact, one and the same and that the re-pairing and renesting of ducks is expedited because of this.

While these late-season or renesting courtship flights of mallards, gadwalls and pintails were common, we seldom saw them in the shovellers and blue-winged teal. I suspect that the difference occurred because of the length of time the drakes stayed with their hens after the clutches were laid. Blue-winged teal and shoveller drakes did not abandon their hens until incubation was well advanced; while mallard, pintail and gadwall drakes abandoned their hens shortly after the clutches were completed. This meant that few unmated hens resulted from early nest destruction among shovellers and blue-winged teal, while unpaired hens of the other species were common.

Retrieving of displaced eggs.—Eggs occasionally rolled out of a nest when the hen left quickly. When she returned and was on her nest again, the hen retrieved these, using the under part of her bill. Retrieving behavior has been described for several species of birds, and this literature is reviewed by Noble and Lehrman (1940:36).

Lorenz and Tinbergen (1937:10-11), describe and illustrate the retrieving of eggs by the grey-lag goose. The retrieving behavior of the grey-lag, as they describe it, is similar to the retrieving behavior I have seen among ducks.

In July, 1949, I experimented with a shoveller hen to see how and whether she would retrieve eggs when they were placed outside her nest. Figures 28 and 29 show the method by which this incubating shoveller hen recovered the displaced eggs. All retrieving was done by the hen from her position on the nest. No attempt was made to retrieve eggs by means other than the underside of the bill, or from distances greater than she could reach as she sat on the nest. I found that eggs which could not be retrieved by this means, or from a distance the hen could reach easily, were abandoned. For example, on July 16 I made the following observation: I had left three eggs in the nest and had put two more eggs ten inches from the nest. On her return, the hen quickly pulled two eggs into the nest, but the third egg rolled away each time it struck a grass clump. After thirteen attempts to bring it in, the hen ceased trying and left the egg where it was. No variations in procedure were attempted.

Lorenz and Tinbergen found that the grey-lag goose retrieved only objects with an unbroken surface. I found that the shoveller hen retrieved only unbroken eggs and that eggshells were carried away. One pintail hen, however, carried away some of the broken eggshells and retrieved others, using the same method as that described for the shoveller, apparently not always differentiating between whole eggs and shells.

Rest and feeding periods.—The time spent away from the nest has been measured by Low (1945:48), who used a recording instrument to indicate frequency and length. He concluded, "No definite rhythm of incubation was detected in redheads as rest periods away from the nest came at irregular intervals during the day and night."

Of the canvasback, Hochbaum (1944:50) says, "I observed no fixed hours of departure from the nest during incubation. Short rest periods during the morning and evening twilight appear to be regular, but the hen also leaves the nest at irregular intervals during the day. Departures become less frequent, and of shorter duration, as incubation advances."

Girard (1941:238) writes of the mallard, "Nesting observations indicated that the incubating hens permitted themselves two feeding periods per day."

Bennett (1938:49) says of the blue-winged teal, "The female left the nest for exercise, rest and feeding purposes once or twice each day. Usually the rest periods were about 7 a.m. and 7 p.m. The females were observed to stay away from the nest 20 minutes to two hours. Occasionally females were seen to leave the nest during the middle of the day."

At Delta, I found that hens leave their nests between seven and eight in the morning (20 observations). Of 30 afternoon observations, none occurred after sundown. My observations of the five species of ducks here treated indicated that hens probably left their nests to feed, stretch, bathe, and preen once or twice each day. The time of day and number of times per day varied with individuals.

The activity of the hen during this off-nest period appeared to be unhurried. I have watched incubating hens feeding along the ditch bank and pothole edges and have had the impression that food was of secondary importance. Most of the off-nest period was spent in preening, bathing, stretching and sitting. Contrasted with the eagerness with which ducks consume food at other seasons, their behavior showed that food consumption at this period was at a minimum. Perhaps the food requirements of an incubating hen are greatly reduced because of her reduced activity.

Egg recognition.—In my experiments, for various reasons, I often placed eggs of one species under a hen of another species. It was found that pintails accepted blue-winged teal eggs, mallard eggs, redhead eggs and shoveller eggs. One mallard hen readily accepted teal eggs, despite their much smaller size. One blue-winged teal accepted mallard eggs, but in smaller number than her own clutch. One pintail hen continued to sit on a mixed clutch containing teal, canvasback, and pintail eggs.

Of five mallards given wooden eggs in April, 1949, all deserted their nests. Of 14 pintails given wooden eggs during the same week, all but two deserted. How much of this desertion would have occurred if the wooden eggs had not been substituted, however, was not determined. All of the mallards and five of the pintails still were laying. One incubating shoveller hen in 1949 was given two wooden eggs. These were added to her clutch of three eggs. This hen buried the two wooden eggs underneath the nest lining and continued to sit on her own three live eggs. The shape of the wooden eggs may have influenced her, as they were considerably longer than her own eggs.

Farley (1939:57) writes of a pintail sitting on stones. In my experiments at Delta, ducks have shown no ability to differentiate their eggs from those of other species and, in many instances, from wooden eggs. The acceptance of dummy eggs, however, did not meet with uniform success, and the substitutes sometimes were rejected.

Eggshell carrying.—An early description of a mallard hen

carrying an eggshell has been given by Oates (1905:33), who believed that the egg had been cracked and that the hen was removing it in an attempt to clean out her nest.

Hochbaum (1944:92) comments on egg-carrying by wild mallard hens as seen by local guides at Delta. He also describes seeing a shoveller hen carrying an egg but concludes, "I have no evidence which might explain such behavior." Lindsey (1946:491) described the New Mexican duck carrying what, from his description, were whole eggs and believed the action to be associated with nervousness and desertion.

I have observed wild hens carrying eggshells on four occasions, and have been able to induce eggshell carrying. When we added shells to nests of wild birds, it was possible to learn the details of this rarely seen behavior. From my observations at Delta I have concluded that these earlier published reports concerned the carrying of eggshells rather than whole eggs.

My first observation of eggshell carrying was on the evening of June 8, 1946. A shoveller hen was seen at a distance of about 40 yards carrying an eggshell.

My second observation of eggshell carrying occurred when Albert Hochbaum and I watched a pintail hen on the evening of June 17, 1946. She was first seen preening along a ditch bank. From all appearances, she was an incubating bird away from her nest for the evening feeding and preening time. She appeared to be almost ready to return to her nest.

After a few minutes she walked slowly up the ditch bank, craned her neck for a careful examination of her surroundings, and then flew west, alighting in the grass 100 yards away. About two minutes later she flew back over the ditch carrying an eggshell with her bill, dropped it over the water, and returned to her place in the meadow from where she had come. We immediately walked over to the spot. As we approached, the hen fluttered away over the grass and dropped down about 60 yards away. At our feet in the whitetop was a partly destroyed nest. Three whole eggs remained; they were warm. Three more eggs had just been destroyed. The tiny wing of the almost-ready-to-hatch embryo lay at the edge of the nest along with matted down and fresh blood. Upon examining the undamaged eggs, we found them to be within a few days of hatching. It seemed likely that the predator, apparently a Franklin ground squirrel, had been at the nest when the hen was away and had been frightened by her return.

The third observation was made on June 3, 1947, at 9:05 a.m. when a pintail hen came over the roadside ditch, dropped an eggshell into the water, and then joined her drake on the ditch bank. She remained there for 30 minutes after which the pair made a flight over the adjacent cover and returned to the ditch bank for another eight minutes. The pair then made a second flight. This time the hen dropped into the cover, and the drake returned to his ditch bank alone. I then walked into the whitetop until, on approaching the nest, I saw the hen fly off with another eggshell. She circled over the ditch, dropped the shell over the water, and, when joined by her drake, flew off to the bay 1,000 yards away. The nest contained no down. One eggshell remained with fresh unincubated contents. The eggs were cold but had been broken recently by what appeared to be a Franklin ground squirrel. This was a new nest. The hen, on her morning visit to deposit an egg, had found her nest destroyed or being destroyed.

A fourth observation was made on June 21, 1947 at 7:00 a.m. when Nina Elder, William Carrick, and I saw a pair of shovellers make several trips over whitetop cover across the road from the Delta ditch. The hen lit in the grass as her drake returned to the ditch bank. One minute later, the hen flew out of the grass with an eggshell. The hen then made three flights back to the nest and each time came out carrying an eggshell. Each time the shell was dropped into the water of the roadside ditch. After almost two hours of watching, we walked into the meadow to examine the nest from which the shoveller had taken the shells. Five whole eggs and one broken egg remained. A Franklin ground squirrel was at the edge of the nest when we approached. It ran into the dense grass as we came nearer. The nest was at the moment being destroyed by the squirrel. The remaining five eggs were fresh and unincubated, showing that the hen apparently was still laying and had been returning to deposit another egg at the time the squirrel was eating the eggs. Since the hen took shells from the nest at the time when the squirrel was destroying the nest, her arrival must have driven off the squirrel temporarily.

Induced eggshell carrying.—The rarity of the above observations becomes apparent when we consider that the approximate number of my full days afield on the breeding grounds each year was over a hundred, and also that the phenomenon was watched for by others. Furthermore, the years 1948 and 1949 passed without yielding a single glimpse of this behavior, albeit more than the usual time was spent afield. During 1947 and 1948, several attempts were made by William Carrick and me to erect a blind near a going nest in the hope of being able to plant shells and of seeing the resultant behavior. These attempts failed because of complete predation.

In July, 1949, a year later, a wild shoveller hen nested within the confines of the large predator-proof hatchery pen at the Delta station. This pen covered about one acre and was surrounded by a high fence, within which was a man-made pond of about one-tenth acre. Thirty yards from the pond where the nest was discovered, a blind was placed; and, on July 5, our first shell-carrying experiment was begun.

The eggs of our hen were in an advanced stage of incubation. We had determined their hatching date to be the ninth of July, or in four days' time. Consistent with the lateness of incubation, the hen was attached firmly to her nest and as tolerant as we could expect any hen to be.

Six fresh shoveller eggs were substituted for her own clutch, which was placed in an incubator for hatching. The purpose, of course, was to induce her to stay on her nest beyond the normal incubation period. Then, five hatched shoveller egg-shells were added to her nest, being placed on top of her whole eggs.

Within 30 minutes of this interruption, our hen returned to the nest. She alighted 20 feet from the nest, walked toward it until she was within 12 feet, then in a "sneaking" posture continued to walk the last six feet. She settled herself on her eggs and eggshells and, crushing them beneath her by a "rock-the-boat" motion, attempted to settle down. One shell would not go into place; grasping this with her bill, she flew away to the pond (Figure 30) and quickly submerged the shell after alighting on the water.

Following this preliminary experiment, we made 13 additional trials. In each trial, shells were added to the nest and the bird's behavior watched and studied. Our hen made 81 trips from nest to pond, each time carrying away a shell.

Further experiments were carried out in 1949 and 1950 to see whether other species of ducks also removed broken eggshells from their nests. On July 12, 1949, Eugene F. Bossenmaier put 10 broken eggshells in a wild blue-winged teal's nest, which was about half-way through incubation. Because no blind had been constructed beside the nest, the hen's activities were watched from a distance by use of binoculars. After 15 minutes had elapsed, the hen returned to her nest. After about one minute she flew off in the direction of the roadside ditch. Seven trips were made by the hen between the nest and ditch during a period of one hour. From our distant observation post we were unable to tell whether or not the hen had an eggshell in her bill. Examination of the nest after one hour, however, showed that no eggshells had been removed by the hen.

On the same day, eggshells were put into a lesser scaup's nest, which also was well into incubation. This hen returned to her nest but left quickly and never returned.

In 1950, I erected blinds beside two pintail nests and added eggshells to the nests in the same manner. One of these hens returned to her nest each time within 15 minutes and, after considerable hesitation and delay, seized an eggshell, paused beside the nest (Figure 31), and flew away with it in the same manner as the shoveller had done. Her flight line took her out of my view, and I was unable to tell exactly where she went. On each occasion, however, she headed for the roadside ditch. Each time that shells were added to the nest, this hen carried away one or two and then sat on the remaining shells on her eggs.

The other pintail hen was wary and returned to her nest cautiously after one hour's absence. She carried out one eggshell and then failed to return. Assuming that she had deserted her nest permanently, I left the blind. Although I discovered her on her nest again the following day, I conducted no further experiments with this bird.

In 1950, I erected a blind beside a wild baldpate nest in the Delta hatchery pen. When I gave this hen eggshells, she deserted her nest.

I have found no evidence that hens remove the eggshells from nests at hatching time as is common among altricial birds. Hatched nests frequently were found with old shells unremoved.

The tendency for ducks to carry eggshells to water probably accounts for the shells which frequently are found in the bottoms of dried-up sloughs in summer.

Response to nest moving.—On the prairies of central Canada, I have heard farmers describe how they save duck nests from destruction by moving them from the path of the plow or cultivator. Many times these changes are successful when ducks are well-advanced into the incubation period.

One such incident occurred in the spring of 1947. Mr. Murry Werbiski, a farmer on the Portage Plains nine miles south of the Delta Marsh, reported a pintail nest which he had moved. This nest had been moved four times. The first time it was moved about seven feet, the width of the cultivator. The second and third times it was moved about the same distance, and the fourth time it was moved about 20 feet. As the hen became accustomed to the tractor, she allowed it to approach to within a few feet before flushing. When I visited the nest on the evening of May 24, after it had been moved four times, it held ten eggs which were advanced in incubation. As we came to the site on foot, she flushed at about 20 feet.

The amount of disturbance that nesting ducks can stand without deserting is variable. It seems to be greatest where disturbance is recurring so that the bird becomes conditioned to it, and where cover is scant.

I tried this practice of moving nests with mallards, pintails, shovellers, and blue-winged teal and found that their nests could be moved successfully. These birds readily accepted the changes when in the later stages of the incubation period. Movement of the unfinished clutches during the laying period invariably led to desertion. Success undoubtedly would depend too, on how far the nest was moved at one time; and it also would vary with individual birds.

It is important to move not only the eggs but also the materials which make up the nest lining in order to have the hen accept her nest in a new place. The reason for this seems to be a conditioning of the bird to the nest site as well as to the eggs. Tinbergen (1951:149) wrote: "When a broody herring gull has to choose between the empty nest and the eggs in an artificial nest about a foot away, it will sit in the empty nest because it is on the accustomed site. The bird will retrieve the eggs, or even occasionally sit on them in an artificial nest, but the attachment to the orginal site, acquired during the building of the nest, is extremely strong."

A strong attachment to the nest site was evident in an experiment I conducted with one shoveller hen. On July 6, 1949, I left three of seven eggs in a shoveller nest and placed four eggs about one foot away from it. As I watched from a blind, I saw the shoveller hen return and sit on the four eggs outside the nest. She remained on these eggs only three minutes, however, and then returned to her original nest site, and while sitting on the three eggs, pulled the outside group of eggs to her from one foot away.

Disappearance of eggs.—During the course of this study, nests of laying and incubating birds were found. When revisited a few days later a nest of a laying hen normally held a larger number of eggs than when first found. Nests of incubating birds normally had the same number of eggs. This, however, was not always the case; and the loss of eggs from an unknown cause was frequent. In 12 instances I found a dwindling number of eggs.

The frequency of egg loss during the incubation period (short of complete destruction) probably is far greater than these data seem to indicate. The eggshell-carrying behavior of the hen doubtless accounts for part of the disappearance of eggs.

SUMMARY

1. On warm evenings following arrival, mallard and pintail pairs cruised the nesting meadows in low circling flights. These flights were curtailed by low temperatures.

2. Cruising flights over nesting meadows preceded egg laying by four or more days.

3. The only part of nest construction which preceded the laying of the first egg was the scraping of the nest bowl.

4. Nesting material was gathered by the hen from her position

on the nest. I found no evidence that surface-feeding ducks ever carried nesting material from distances beyond their reach from the nest.

5. When located in heavy grass, the surface-feeding ducks pulled surrounding vegetation over their bodies to form a concealing canopy over the nest.

6. Normally, one egg was deposited by the hen each day until the clutch was completed.

7. Laying mallard, pintail, and gadwall hens usually went to the nesting meadows during the first four hours of daylight. Shoveller and blue-winged teal hens went to the meadows to lay later in the morning.

8. The drakes accompanied their laying hens to the nesting meadow; each waited not far from his hen at a loafing place and often made low flights over the meadow where she was located on her nest.

9. Mallard, pintail, and gadwall drakes abandoned their hens shortly after the clutches were completed. Shoveller and blue-winged teal drakes remained paired with their hens later into the incubation period.

10. Like many other species of birds, ducks were found to sit for long periods past the normal incubation time on dead or infertile eggs before abandoning them. One pintail hen was known to sit on the same nest for 60 days, or 39 days past the regular incubation period.

11. Only occasional prenuptial courtship of surface-feeding ducks was seen at Delta because most of the birds were paired when they arrived and were in their nuptial period. Hence, during the first three weeks of the nesting season, little courtship activity was seen.

12. Late in the nesting season, "teasing flights" of renesting courtship of mallards, pintails, and gadwalls were common. The incitement call of this courtship was induced in incubating hens by disturbing them and keeping them off their nests.

13. The emitting of the incitement call on disturbance seemed to be the first stage of courtship leading to re-pairing and renesting. It was not common in shovellers and blue-winged teal. The drakes of these species remained with their hens far into the incubation period; and, for this reason, unpaired hens were less common in these species than in mallards, pintails, and gadwalls.

14. It was found that shoveller and pintail hens retrieved eggs

from outside the nest by pulling them into the nest with the under side of their bills. They were not always able to differentiate between whole eggs and broken eggs.

15. Broken eggshells were removed from their nests by hens. The shells were carried with the bill and taken to water. One shoveller hen pushed each shell under the water and waited on the water until the shell sank.

16. The tendency to carry shells to water apparently accounted for the eggshells commonly found on the bottom of dried-up sloughs and ditches in late summer.

17. Eggs often disappeared from nests during the incubation period. These eggs undoubtedly were removed by the hen following breakage by the hen or by a predator.

18. Some ducks will allow their nests to be moved during the incubation period without deserting them. It was a common practice among Manitoba farmers to move nests found in fields during farm operations.

CHAPTER VIII

BREEDING SEASON MORTALITY

DUCKS, like most other ground nesting birds, have high reproductive potentials, but are vulnerable to many forces. Fire, flooding, agricultural practices, and a variety of predators are decimating factors pulling downward against the high reproductive rate. Although losses are occurring throughout the year, the nesting season losses probably are much higher than the natural losses at any other season, with the possible exception of those caused by botulism, of which great numbers of ducks die.

In the Delta marsh, losses of nests were caused by crows, skunks, Franklin ground squirrels, agricultural practices, fire, flooding and trampling of cattle. After the young were hatched, the old enemies became less important, but new ones were added to the list. The mink, marsh hawk, Franklin ground squirrel, great horned owl, crow, violent storms and drought were potential enemies of the young ducklings. Adult mortality was caused by predators surprising hens on the nest and by accidental deaths.

Thus, three categories of mortality occurred on the Delta marsh and probably on the many other marshes within the range of breeding waterfowl. These types of losses can be classed as: (1) loss of nests, (2) loss of young, and (3) death of adults. Of these three, the first is the easiest to study.

Nest Mortality.—The destruction of duck nests (Figure 32) by predators was studied intensively by waterfowl investi-

113

gators in several areas between 1935 and 1945. Earliest of the projects was that of Kalmbach (1937) who worked on crow-waterfowl relationships in Alberta and Saskatchewan. Kalmbach found that, of 512 duck nests, 270, or 52 per cent, were destroyed before hatching. The crow led the list of predators and took 31 per cent of the nests.

In Saskatchewan, Furniss (1938:24) reported on 108 nests. Of this group 25 per cent were destroyed.

In western United States, Williams and Marshall (1938:-41-43) reported the fate of 1,560 nests and found that in the Bear River marshes of Utah predators were of small importance, taking fewer than 7 per cent of the eggs.

In Montana, Girard (1941:242) studied the nesting success of mallards and discovered that of 1,797 eggs, 71.2 per cent hatched and 20.4 per cent were destroyed by predators. The crow took only 5.6 per cent.

Girard (1939:368) also studied nests of the shoveller and found that predators destroyed 23 per cent of the 1,135 eggs under observation.

In regions nearer Delta, Kalmbach (1938:614) reported the success of 917 duck nests on the Lower Souris Refuge in North Dakota, studied in 1936 and 1937. In addition to the 917 nests studied, Kalmbach (1939:597) had available to him the records of 620 duck nests studied by M. C. Hammond. In reviewing all nesting success studies in this later paper, Kalmbach showed that of the entire group of nests studied on the Lower Souris Refuge (1,537 nests) 64 per cent were destroyed by predation.

Nest predation studies were made at Delta in 1939 and 1940. They showed that the Delta marsh region had relatively low nesting success and heavy nest mortality. Of 206 nests studied (Sowls 1948:130) only 35 per cent hatched successfully. Crows destroyed 21 per cent of the nests, 10 per cent were deserted, 8 per cent were flooded, skunks took 8 per cent, and the cause of destruction of 6 per cent was unknown.

Between 1946 and 1951, no nesting success studies were made at Delta. In the renesting study, however, it was apparent that the predation was little different from that in 1939 and 1940.

Figure 32. The appearance of these mallard eggshells indicates
crow destruction.

Figures 33 and 34. Slowly rising water sometimes causes hens to build up nests. This pintail nest had a top layer of four eggs and a bottom covered layer containing two eggs.

Human disturbance and nest predation.—The techniques employed in these nesting-success studies were generally the same. That is, the nests were found by searching and were revisited to see whether they hatched or were destroyed. This technique introduced the possibility that intrusion by human beings may have helped predators in finding nests. This danger is great and must be recognized. There is, however, no way of determining the extent to which human intrusion aids predators.

This subject has been treated in detail by Kalmbach (1937: 6-7) with special reference to the crow. Kalmbach says:

"The writer is convinced that, when the incubating bird is not actually flushed in the presence of crows, or when its eggs are well covered and the nest is left by the observer reasonably well concealed, his intrusion will not afford the crow an important clue as to nest site.

"It cannot be emphasized too strongly, however, that careless intrusion of human beings into duck-nesting areas creates a hazard of utmost importance, for incubating ducks may then be flushed in the presence of crows and the suddenly uncovered eggs left exposed to view. It is for this reason that the breeding grounds of ducks should be carefully guarded against trespass during the nesting season."

Where the main predators are skunks and Franklin ground squirrels, I believe that intrusion is an important factor in increasing predation, and that the number of nests destroyed by these two mammals at Delta may, therefore, have been increased by our disturbance.

Early in spring, when dead vegetation makes up the cover of the nesting meadows, a man can traverse the ground without leaving a conspicuous trail behind him. But in summer, when grasses are green and succulent plant growth (such as sow thistle) is abundant, a man's trail remains as he moves from nest to nest, and it seems likely that such a trail could be followed easily by a skunk or squirrel.

Flooding as a cause of nest loss.—Heavy rains or strong winds blowing off a large body of water, such as a lake or bay, cause heavy loss of nests by flooding. At Delta, the pintail, of all the surface-feeding species, has been found to suffer the heaviest loss from rising waters. In 1939 and 1940, the nesting outcome of 34 pintail nests was observed. Of this number, 26 per cent were destroyed by flooding. The normal rainfall for June in southern Manitoba is about 2.5 inches. In 1940, 4.74 inches of rain fell in June.

The pintail falls easy victim to rising waters because of its tendency to nest in small hollow areas. All of the 1940 nests of pintails that were flooded by rains were located in slight depressions on high ground. During these studies the pintail was the only species found to nest often in depressions. Of the pintail nests studied, seven were found in abandoned Richardson ground squirrel burrows which had been enlarged at the entrance by some larger animal.

Destruction of nests by flooding.—Nests located in areas where waters rise slowly often are built up by the hen as the water advances. Additional eggs may be laid, and some of the eggs already in the nest at the time of flooding may be brought up with the addition of new nesting material. I have found this behavior common in mallards and pintails but never have observed it among other species of surface-feeding ducks. Figures 33 and 34 show two levels of a built-up pintail nest. In Figure 33 the entire nest with four eggs is shown as it appeared when found on May 10, 1946. The water beneath the nest was four inches deep, and the top of the nest was eight inches above water. Figure 34 shows a lower level of the same nest where two eggs had been buried by the addition of new nesting material.

The partial destruction of puddle duck nests by flooding, resulting in a reduced clutch of eggs, was observed by me on two other occasions. One of these nests belonged to a pintail and the other to a mallard.

Adult mortality.—No one can know how many adult ducks die during the breeding season. During the period of this study, however, some general impressions of the mortality of adults were obtained. While we were hunting for nests on the study area during 1946, 1947, 1948, 1949, and 1950, a total of 15 dead adult ducks were found. Eight of these were hens which had been killed on their nests by unidentified predators. Of this group, five were blue-winged teal, two were pintails, and one was a mallard. Of the seven drakes found dead, four apparently were killed by striking telephone wires which

ran the entire length of the study area. Three were killed by unidentified predators. In addition to the drakes found under the wire, four dead coots were found. These obviously had been killed in the same manner. No hens were found dead under the telephone wires during this study; possibly due to their familiarity with the area.

Although the number of hens killed on their nests was comparatively small, it may, in some areas, help account for the well-known excess of drakes that exists in waterfowl.

Food habits of marsh predators.—Some understanding of the relationship between waterfowl and predators can be obtained by an analysis of the food habits of the predators. To learn the extent to which our predators preyed upon ducks, their young and their eggs, we gathered data on nest predation at Delta in 1939 and 1940. During that time, collections and analyses of adult and juvenile crow stomachs, ground squirrel stomachs, marsh hawk pellets, mink scats, and the remains from marsh hawk nests also were made. Data, discussions and conclusions concerning these analyses are given by species.

Foods of adult crows.—The analysis of stomach contents of 66 crows collected during the duck nesting and brood seasons are given in Table 15. From these data we see that the crow on the summer waterfowl breeding ground is an omnivorous bird. Frogs, toads, miscellaneous insects and a variety of seeds were the most common items consumed. It is apparent that the infrequent occurrence of eggshell remains (four times out of 66) had little significance as an indicator of crow damage to duck nests. The presence or absence of eggshells in the stomachs of crows depends largely upon which crows (breeders or wanderers) are shot, and whether or not the preying crows consumed shells when eating eggs. We would expect the wandering, immature crows to consume many more eggs than the resident, mature birds. Even the habitual nest robbers might not have eggshell remains in their stomachs if, in taking eggs, they swallowed few shells. Since duck egg contents leave no trace in the stomachs of crows, the consumption of duck eggs is not indicated in the table.

Foods of nestling crows.—Fifty-five nestling crows were collected and their crops analyzed. The results of this analysis are given in Table 16. The food of the nestlings was carried to them by the parent crows, and it showed a similar variety. Two conspicious differences were evidenced: (1) remains of ducklings were found four times in nestling crow stomachs; (2) farm grains occurred in the stomachs of adult crows but not in the stomachs of nestling crows. Whether or not the first of these differences was significant could not be determined. The presence of farm grains in some adult crow stomachs probably is due to the extensive feeding areas of wandering crows and the limited feeding areas of parent crows. Since both nestling and adult crows were collected on the lake ridge several miles from grain fields, and since the ridge was used by breeders and non-breeders alike, it seemed likely that the non-breeders were the grain-eaters and that nestling crows and parent crows were

TABLE 15. FREQUENCY OF OCCURRENCE OF ITEMS FOUND IN SIXTY-SIX ADULT CROW STOMACHS COLLECTED AT DELTA, MANITOBA IN 1939

Item	Number of Times Occurring	Item	Number of Times Occurring
Bird Remains		Amphibian Remains	
Passerine	2	Frog and Toad	22
Eggshell	4	Seed Remains	
Mammal Remains		Wheat	3
Mice	1	Oats	2
Insect Remains		Barley	4
Beetle	18	Cherry	1
Fly	1	Poison Ivy	3
Grasshopper	1	Dogwood	1
Dragon Fly	1	Unidentified Vegetation	5
Unidentified Larvae	3	Empty	10
Unidentified Pupae	1		

not. This is consistent with the wandering habit of non-breeding crows which will be discussed later.

Foods of marsh hawks.—The contents of 53 marsh hawk pellets collected around marsh hawk nests is given in Table 17. The numbers of times food items were found at the same nest are given in Table 18. From these data, it is apparent that ducklings as well as adult and juvenile coots occasionally are taken by marsh hawks. It is apparent, too, that passerine birds, principally juveniles, and several species of rodents make up an important part of the marsh hawk's diet.

Foods of mink.—Bennett (1938:70) found that practically every fecal dropping of mink collected between June 15 and July 15 contained remains of coot chicks. Low (1945:59) found the same to be true in his studies. Both Bennett and Low concluded that the coot acts as an important buffer between the mink and ducklings. Our information collected at Delta substantiates the findings of these two workers and is given in Table 19. Coots, muskrats and small rodents doubtless act as important buffer species in cutting down the loss of ducks to mink. Where these buffer species are scarce, the mink may be an important predator on waterfowl.

TABLE 16. FREQUENCY OF OCCURRENCE OF ITEMS FOUND IN FIFTY-FIVE NESTLING CROW STOMACHS COLLECTED AT DELTA, MANITOBA IN 1938 AND 1939

Item	Number of Times Occurring	Item	Number of Times Occurring
Bird Remains		Insect Remains	
Diving Duck	1	Beetle	25
Unidentified Duckling	3	Grasshopper	1
Passerine Nestling	3	Cricket	1
Unidentified Passerine	4	Unidentified larvae	3
Eggshell	3	Unidentified pupae	3
Mammal Remains		Amphibian Remains	21
White-footed Deer Mouse	1	Cherry Stones	3
Shrew	1	Unidentified Vegetation	7
Unidentified Mouse	3	Empty	2

Franklin ground squirrel foods.—The foods taken by this rodent at Delta have been reported by Sowls (1948:123). Although this squirrel was found to be an important predator, and was capable of killing young ducklings, no duckling remains were found in 178 stomachs.

Changes in predator densities.—It is well known that predator numbers sometimes fluctuate violently from one year to

TABLE 17. FREQUENCY OF OCCURRENCE OF ITEMS IN FIFTY-THREE MARSH HAWK PELLETS COLLECTED AT DELTA, MANITOBA IN 1939

Item	Number of Times Occurring	Item	Number of Times Occurring
Bird Remains		Richardson Ground Squirrel	1
Unidentified Passerine	20	Snowshoe Hare	1
Juvenile River Duck	3	Red-backed Mouse	4
Adult Coot	5	Meadow Mouse	13
Juvenile Coot	1	Insect Remains	
Juvenile Blackbird	1	Grasshopper	1
Mammal Remains		Unidentified Pupae	1
Franklin Ground Squirrel	2		

the next. At Delta, both in earlier studies and in the renesting study, various levels of predator densities have been apparent. In the years 1939 and 1940, heavy populations of Franklin ground squirrels existed in the wooded ridge bordering the duck nesting meadows. During the years 1946 through 1950, the density of this predator had diminished materially but was great enough to handicap the conducting of the renesting study.

Skunks, which were scarce during 1939 and 1940, were abundant during the years 1946 through 1950. Trapping and shooting of skunks on the study area throughout this period did not seem to lower the skunk population, although an aver-

TABLE 18. FOODS FOUND AT A MARSH HAWK NEST, DELTA, MANITOBA, 1939

Item	Number of Times Occurring	Item	Number of Times Occurring
Bird Remains		Unidentified Adult	
Juvenile Coot	1	Passerine	2
Sora Rail	2	Mammal Remains	
Unidentified Passerine		Meadow Mouse	3
Fledgling	2	Jumping Mouse	1
Juvenile Redwinged		Deer Mouse	2
Blackbird	4	Unidentified Mouse	2
Adult Yellow-headed and		Young Snowshoe Hare	1
Red-winged Blackbird	4	Richardson Ground Squirrel	1

TABLE 19. FREQUENCY OF OCCURRENCE OF ITEMS FOUND IN 122 MINK SCATS AT DELTA, MANITOBA IN 1938 AND 1939

Item	Number of Times Occurring 1938	1939	Item	Number of Times Occurring 1938	1939
Bird Remains			Mammal Remains		
River Duck	15	6	Muskrat	12	2
Diving Duck	5	2	Mouse	3	22
Unidentified Duck[1]	11	6	Franklin Ground		
Coot Adult	3	6	Squirrel	0	3
Coot Juvenile	1	2	Snowshoe Hare	9	7
Grebe	0	1	Amphibians	3	2
Passerine	6	8	Insects	3	8
Unidentified Bird	3	5	Unidentified Contents	4	1
Eggshell	0	1			

[1] Of the duck remains 3 were adult, 16 juvenile, and 12 unidentified in 1938; 5 adult, 6 juvenile, and 3 unidentified in 1939.

age of about 25 skunks per year were taken on the 3,115-acre study area during this period. There appeared to be a constant movement of skunks into the area, which made limited control futile.

One of the greatest changes in predator densities, which has been apparent over a long period of time in the prairie provinces of Canada, has been the northward extension of the range of the eastern crow. Considerable historical information exists to substantiate this recent advance following the northward spread of agriculture. Phillips (1928) says, "The crow has increased rapidly during the past few years in the agricultural districts."

Of the crow in Alberta, Farley (1932:47) says, "Two outstanding changes of economic importance have taken place in the bird-life of Alberta during recent years, viz: the extraordinary increase in the number of crows and the deplorable decrease of many species of our wild ducks. The crow, like its congenitor the magpie, is a comparatively new addition to the avifauna of Central Alberta, although both birds were here in the buffalo days. In 1892, the year of my arrival in Alberta, I did not record a single crow in the list of birds observed that season. Only the odd pair was seen during the next few years, and it was not until the beginning of the present century that their numbers showed any material increase. By 1910 there were evidences that crows would shortly become a menace to our water-fowl should their increase continue as it had in the previous decade. And, unfortunately what was expected did happen. The increase has been phenomenal and there appears to be no satisfactory explanation for the extraordinary change of status."

Kalmbach (1937:7) says of the crow in Alberta and Saskatchewan, "The crow is partial to areas devoted to agriculture and to a large extent is now dependent on them. In contrast to the raven, which has receded with the advance of agriculture, the crow has extended its range in the north to the limits of such developments."

For the Delta region, Hochbaum (1944:11) says, "The crow had increased since the prairie was broken for agriculture—old-timers remember when it was an uncommon bird in the Delta region."

The effects of predator population densities upon the percentage of duck nests taken is difficult to determine. We are not certain, for example, that six skunks in a nesting meadow do more damage than one. This would depend largely upon the vulnerability of the nests and the movement and hunting patterns of the predators. Too, it would depend upon how much effect one predator might have in the absence of other predators. Obviously, a crow cannot destroy a nest when it already has been destroyed by a skunk.

Errington (1946) speaks of "intercompensations" and describes the increase of pressure by one predator as another predator becomes less abundant. In most waterfowl breeding marshes several predators are acting upon duck nesting populations at one time. At Delta, crows, Franklin ground squirrels, and skunks were competing for the same nests.

Movements of predators.—During the first half of each nesting season, damage by the Franklin ground squirrel always was less than during the latter half of the nesting season. This apparently was due to two factors: (1) the presence of water barriers early in the nesting season, and (2) increased movements of ground squirrels later in the season. Much of the finest nesting cover was immune to ground squirrel attack for long periods in spring because of a moat of water which these mammals could not cross readily.

Unlike the ground squirrel, the skunk did not wait until the meadows dried up before moving into the duck nesting areas. On many occasions it was apparent that skunks lived in the nesting meadows at Delta for long periods. In place of ground burrows in which to hide and sleep, lone skunks built grass beds in the marsh. On May 10, 1946, Peter Ward and I found a grass bed 10 inches high and 2 feet square containing a resting skunk. The ground was too wet for burrowing. There was no evidence that the grass form first had been made or used by a bird or other mammal.

The movement of crows over nesting meadows was of two kinds. One was the movement of crows which were nesting near the meadows. Destruction by these nesters was limited to an area near the wooded ridges and prairie bluffs where their nests were located. The second was made by a more destructive group of crows which were present over the marshes in wandering bands. These crows were seen moving great distances over the marsh, often in groups of a half-dozen or more.

In regard to the movements of different groups of crows and the damage done by each group, Good (letter dated January 16, 1951) says of crows he has observed in Ohio and Manitoba:

"First of all it is definitely established that crows do not breed until the second spring after hatching. These non-breeders are distributed in two groups here in the midwest. First, many return to the woodlot where they were hatched and hang around the nests of crows breeding there (possibly or probably their parents). Others remain in small bands feeding across the countryside and roosting in small roosts (up to several hundred birds) at night. While we were at Delta these summer roosts were observed and small bands of crows which were most certainly non-breeding birds were feeding together. Little evidence of this kind was noted at Delta but several such observations were made in the Minnedosa region.

"As for the breeding birds, we watched several nests containing young and found the feeding habits of the adults to be remarkably like those in Ohio. The adult birds were solitary on these expeditions from the nest, and did their food gathering in open fields or in short grass or other vegetation low enough to permit them to walk about freely. We never saw them hunting over the marsh or in good duck nesting territory.

"On one or two occasions I watched lone birds hunting over the marsh. I could not determine the age or status of these birds but they were obviously hunting. The characteristic flight with the beak pointed down is good evidence. The birds observed beat slowly into the wind and when nearly across the marshy area would catch the wind and allow themselves to be quickly carried back to the far side where they would begin again to beat slowly into the wind on a new path across the marsh."

Predator control.—Various programs of predator control have been advocated for the waterfowl breeding grounds. Most of these have been concerned with control of crows. For example, Phillips (1928) says, "The Alberta game department has at present in operation a competitive prize system whereby the game clubs and other organizations compete for the largest number of crows and magpies killed and turned in for examination." Phillips gives the data for the number of crow and magpie eggs and adults turned in for the 4-year period beginning in 1924. The number of eggs turned in during these years ranged from 44,769 to 107,116 and the number of adult crows and magpies ranged from 22,275 to 44,652. Similar programs on a larger scale have been sponsored since by Ducks Unlimited, a private organization financed by donations from duck hunters, and put into practice through the cooperation of gun clubs, schools and sportsmen's groups. These programs have included the paying of bounties for adult crows and also for crow eggs and crow nestlings. It is conceivable that these practices may be valueless, or may aggravate rather than alleviate crow predation on ducks nests. In view of Good's discussion of crow movements, it seems plausible that the destruction of crow nests may add more wandering adult crows to our marshes, these being more destructive of duck nests than the comparatively sedentary breeding adult crows.

Herein lies one of the unsolved problems of the crow-waterfowl relationship. Kalmbach (1937) demonstrated the destructiveness of the crow to nesting ducks. The historical record is well substantiated in showing that the crow is new to many parts of the continent's waterfowl breeding range. New research should be initiated to determine the advisability of robbing crow nests of eggs and young.

In regard to control of the Franklin ground squirrel the conclusions were: (1) "ground squirrel predation upon duck nests is a local problem in certain places at certain times," and (2) "ground squirrel control in wild marsh areas is too costly and too harmful to beneficial forms of wildlife to be practical. When populations are dense enough, disease and parasites cause a sharp decline in numbers. Artificial control may do more harm than good."

No large-scale skunk control was carried out at Delta. Effects of skunk control upon nesting success of ducks have been studied by Kalmbach (1938:614). He says:

"During the two years of this study on the Lower Souris Refuge depredations by skunks constituted the outstanding factor affecting the welfare of duck nests and their eggs. In 1936 this element of loss (30.4 per cent) was almost equivalent to that inflicted by crows as recorded in the Canadian studies. In 1937, after the removal of 423 skunks by trapping on portions of the refuge, a loss due to skunks of only 6.4 per cent was noted. The removal of these animals was done for the definite purpose of learning experimentally what effect such trapping would have on waterfowl production.

"Another factor having a possible bearing on the lowered skunk pressure in 1937 was the fact that two expansive areas, known respectively as Ison Island and Newburgh Island, which harbored the densest concentrations both of nesting ducks and skunks in 1936, were only sparsely populated in 1937. Skunks destroyed 57 per cent of the nests in those areas in 1936, yet, although few or none of these mammals were removed from them in the program of control during the winter of 1936-1937, in the latter year there was only an 8.3 per cent loss from this cause. One cannot, therefore, evaluate fully the benefits of this skunk control until more nearly comparable areas and similar conditions of duck-nest distribution present themselves or until the results obtained over a period of years are appraised."

Evidence that large numbers of one predator may make up for lower densities of another seems to be suggested in Kalmbach's findings. Where crows were abundant on the Canadian marshes studied by him, and skunks scarce, the loss from crows was 31.0 per cent and there was no loss from skunks. Where losses from skunks on the Lower Souris Refuge amounted to 30.4 per cent, loss from crows was only 1.7 per cent.

In any predator-control program designed to cut down nesting losses of waterfowl, it appears, therefore, to be useless to control one predator where several are present. Methods to control all the predators in our wild waterfowl marshes have so far been found to be ineffective and costly. Furthermore, the role of nest predation in waterfowl production seems less significant now than it did during the years when the nesting-loss studies were made. We now know that renesting is a compensating factor for early season predator loss. This will be discussed in the next chapter.

SUMMARY

1. Several nesting success studies were reviewed. The 1939-40 nest predation study at Delta showed a heavy nest mortality. This study showed that predation had not changed materially in 1950 from what it had been in 1939-40.

2. Human disturbance, as a factor in nest predation, is of considerable importance.

3. Flooding of nests causes considerable nest loss. In a year of heavier than usual rainfall in June, a flooding loss of pintail nests was 26 per cent.

4. Slowly rising water sometimes causes mallards and pintails to build up their nests.

5. Of 15 dead adult breeding ducks found, eight were hens killed by predators, four were drakes apparently killed by striking telephone wires, and three were drakes killed by unidentified predators.

6. It is suggested that the mortality of hens on the nest may help account for the well-known excess of drakes among waterfowl.

7. A food-habits study was made of adult and juvenile crow stomachs, ground squirrel stomachs, marsh hawk pellets, mink scats, and the remains from marsh hawk nests.

8. On the waterfowl breeding grounds the crow was an omniv-

orous bird. Since the contents of duck eggs leave no trace in a crow's stomach the data gathered did not indicate the full destruction caused by this predator.

9. The study of 53 marsh hawk pellets showed that the marsh hawk preyed upon ducklings and coots, as well as upon passerine birds, mammals, and insects, but that the last three were the most important.

10. An examination of 122 mink scats showed that minks ate ducks and ducklings along with other birds, mammals, amphibians and insects. Coots, muskrats and small rodents, no doubt, act as important buffer species. When these buffer species are at a low population level, the mink may be an important predator on waterfowl.

11. In 178 Franklin ground squirrel stomachs no duckling remains were found. However, in an earlier study it was pointed out that the Franklin ground squirrels are important predators and capable of killing young ducklings.

12. Franklin ground squirrel and skunk populations fluctuated greatly during the period of these studies.

13. There has been a northward extension of the range of the eastern crow following settlement of the country.

14. Crows do not breed until the second spring. The non-breeding juveniles probably are the most destructive to duck nests because they wander long distances over marshes, while breeding pairs restrict their hunting to the vicinity of the nest.

15. In many marshes, there are several predators competing for duck nests. When one predator is reduced, the damage from another one is increased. Unless waterfowl predator control can be complete and exhaustive, it seems advisable to consider with extreme caution the partial destruction of predators.

CHAPTER IX

RENESTING

DUCKS are "indeterminate" layers, capable of producing large clutches of eggs, as contrasted with the "determinate" layers such as doves, pigeons, and plovers where clutch size is a nearly predictable number. While those birds usually have more than one brood per year, ducks have but one. However, ducks are persistent renesters and will try again if their first nest is destroyed.

While the extent of nest losses among ducks has been studied thoroughly, the details of the renesting phenomenon in waterfowl has received little attention. Studies of renesting in waterfowl were started at Delta in 1946 and continued through 1950. The preliminary findings of this study have been published (Sowls, 1949). This chapter is a continuation of the earlier paper with the same procedures extended to a larger sample size. Work since 1949 also was directed toward the problem of renesting following the loss of downy young.

Previous work.—Bennett (1938:57-58) and Low (1945:50) have distinguished renesting attempts from first nests by smaller clutch size, less down and the general appearance of the nest, and the lateness of the season. In predation studies, Kalmbach (1937:21) classified the nests into "early" and "late season" nests, the latter group apparently representing renests. Cartwright (1944:327) and Errington (1942:170-171) discussed some of the theoretical aspects of the importance of renesting in game birds, including waterfowl. Hochbaum (1944:158) pointed out the need for renesting information about ducks and suggested nest-trapping and marking of hens as a means of study.

Barnes (1948:449) described the occurrence of a banded wood duck nesting twice in one season. Engeling (1949:6) said of a mottled duck in Texas, "I observed a pair build five nests in succession and lay a total of 34 eggs before finally hatching a brood of nine."

During the year 1950, I concentrated on the behavior of hens after nests were destroyed during the *later stages* of incubation in order that this aspect of the problem could be treated more thoroughly.

Clutch size, first nests versus renests.—Lack (1947) pointed out that the tendency for clutches of renests to be smaller than clutches of first nests among ducks has been known for many years in Europe. I already have mentioned Bennett's and Low's work on clutch size.

My data on clutch size of both first nests and renests for 21 individual hens are given in Table 20. On analysis, the apparent drop in clutch size was found to be statistically significant.

Although the average number of eggs in first clutches is greater statistically than the average number of eggs in renests, the difference is not great enough to distinguish first clutches from renests.

Some known renests of gadwalls, blue-winged teal and pintails had more eggs than the first clutch of another individual of the same species. One pintail had the same size clutch in her first nest and her renest.

In Table 21, I have divided the nesting season into three arbitrary parts and have included under each part of the nesting season, all clutches known to be full sets of eggs. It is apparent that a decrease in size occurs in all species as the season advances.

Figures 35, 36. First and second nests of pintail hen No. 47-604106 for 1948.

Figure 37. Third nest of pintail hen No. 47-604106 for 1948.

TABLE 20. CLUTCH SIZE OF FIRST NESTS AND RENESTS OF INDIVIDUAL HENS

Species	Number Eggs First Nest	Number Eggs Second Nest	Number Eggs Third Nest
Mallard	10	9	
Pintail	10	9	
	10	8	8
	10	7	
	9	8	7
	9	7	
	9	6	
	9	4	
	8	8	
	8	7	
	8	7	
	8	6	
	7	6	
	7	3	
Gadwall	11	9	
	10	9	
	8	5	
Shoveller	12	9	
	12	8	8
B-w Teal	12	10	
	10	9	

Bennett's "normal" nests of the blue-winged teal averaged only 9.3 eggs. One blue-winged teal at Delta in 1948 had more eggs (10) in her second nest than Bennett's average for all first nests. The other teal in this study had about the same number of eggs (9) in her second nest as Bennett's average first clutch in a species where clutches of 12 or more are common. The above instances may indicate that many of what Bennett called "normal" nests actually were renests.

The accumulated evidence appears to show that clutch size is not a valid criterion for distinguishing first nests from renests. Many renests would fall into the category of first nests if judged on the basis of clutch size alone. Low (1945:50) recognized this in his study of the redhead and suggests that

TABLE 21. DIFFERENCES IN CLUTCH SIZE IN THE EARLY, MIDDLE AND LATE PERIODS OF THE NESTING SEASON

Species	Completed By May 15		Completed May 15-June 15		Completed after June 15	
	No.	Ave.	No.	Ave.	No.	Ave.
Mallard	23	10.0	25	8.3	3	9.1
Pintail	45	9.0	46	7.1	14	7.0
Gadwall	..	...	17	10.5	10	9.5
Shoveller	..	...	15	10.8	14	8.5
B-w Teal	..	...	54	10.6	42	8.8

his classification of renests does not include all of the renests but only those of which he was certain.

There are other complicating factors which may be inherent in all duck-nesting studies. For example, the number of eggs in a nest may not represent the actual number laid by the hen. A reduction of the clutch occurs when (1) a predator removes an egg, (2) the hen removes an egg broken by a predator, and (3) the hen is disturbed at laying time and drops an egg away from the nest.

Appearance, first nests versus renests.—Some renests were as well constructed and as well concealed as first nests. In some cases there was no noticeable change in appearance between first nests and renests. In two or three gadwall nests there was as much down in the second nest as in the first. In the case of both blue-winged teal nests and mallard nests I could distinguish some reduction of down, but it was not great. In the case of pintails, the species for which I have the most data, there was a noticeable decrease in the amount of down in some renests when compared with the first nests of the same individuals. However, without having seen and recorded the first nest, it would have been impossible to say whether or not the nests were first attempts or renests (Figures 35, 36 and 37).

One of the difficulties in distinguishing renests by appearance is that in order to do so one would have to compare all nests at the same stage of incubation, because down and other materials accumulate gradually.

Renesting interval.—The time between the destruction of

the first nest and the laying of the first egg in the second nest might well be called the *renesting interval*. That this time varies with the stage of incubation has been described for other groups of birds than waterfowl. Stieve (1918) studied the anatomical basis of the re-laying capacity of the jackdaw in Poland. Laven (1940:131) discussed Stieve's findings in relation to renesting. Laven says that when the first clutch is destroyed shortly after the first laying period, re-laying occurs more promptly than when the first clutch is well advanced in incubation. The remaining follicles after cessation of laying soon are resorbed. For a time, however, the larger follicles are capable of being rebuilt. Later in incubation, they are so regressed that they can no longer be used for replacement purposes, and thereafter the smaller follicles must go through a construction process. Accordingly, re-laying takes longer.

For the ringed plover, a two-brooded species, Laven (p. 132) in Germany found an increase in the length of this period which I have called the renesting interval after nests were destroyed, but also found a shortening of the interval again in the very late stages of incubation. When the stage of incubation of the first clutch was 6-12 days, the renesting interval was 5-6 days. When incubation was advanced to 14-21 days, the interval was 11-20 days. When a clutch was 26 days incubated, however, the interval was only 6-7 days. Laven points out that this indicates a renewed regular growth of the follicles toward the end of the incubation period in this species, which normally raises two broods a year.

Not all groups of birds show a lengthening of the renesting interval as the incubation-stage-at-destruction advances. Nice (1937:111) found that the renesting interval in song sparrows usually is five days. The incubation period of this species is 11 to 12 days, and it normally lays three or more clutches a year.

Data gathered at Delta on the renesting interval between 24 first and second duck nests are plotted graphically in Figure 38.

All hens waited at least three days before renesting. For each additional day of incubation at the time of destruction of first nest, these hens waited an average of 0.62 days before beginning to lay their second clutches.

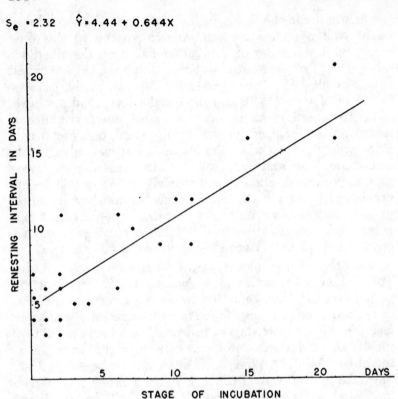

$S_e = 2.32$ $\hat{Y} = 4.44 + 0.644X$

Figure 38. Regression of length of renesting interval on stage of incubation at destruction of first nest.

Continuous laying.—All data so far presented have dealt with the renesting which follows the destruction of a complete clutch after incubation has begun. When nests are destroyed during the laying period, another situation exists.

Information on this phase of renesting behavior has been more difficult to obtain because hens are hard to capture and mark during the laying time. When laying, a hen's attachment to her nest is weaker than it is during incubation, and disturbance often causes desertion. The robbing of a nest each day to determine how many eggs a hen will lay continuously has not been successful in the field because of heavy predation which terminated experiments in all cases.

In 1949, three records were obtained of continuous laying in the wild. A blue-winged teal hen was captured and marked on May 24 after she had laid five eggs. One egg was warm when I found the nest. The eggs were taken from her at the time. On June 2, the same hen was found on a new nest 50 yards from the first nest site. At that time she had nine eggs, but continued laying until she had a clutch of 13. The number of eggs found on June 2 indicated that the hen had continued laying in her new nest the day following the destruction of her first nest. Thus, a total of 18 eggs was laid continuously.

In 1949 also, continuous laying by shovellers was noted twice. On May 20, a banded hen was flushed from her nest where she had laid one egg. She abandoned her nest, moved 80 yards, and began a new nest, this time laying 12 eggs. The first egg in this latter nest was laid the day following abandonment of the first nest.

Incidentally, this individual hen, which was well known to me, nested for four consecutive years on the study area. All four of her first nests, each of which had a complete clutch, held 12 eggs.

The other record of a shoveller laying continuously in 1949 was of a banded hen found on June 27 with four fresh eggs. She abandoned her nest then and continued to lay in another nest 150 yards away on June 28. This time she produced a clutch of six eggs. This late-season nest probably was a second renest, since the clutch was small.

It seems likely from these observations that continuous laying usually occurs when nests are broken up during the laying period, and that there is no renesting interval then.

Further observations concerning continuous laying were made by autopsy. The reproductive system of a laying hen is shown in Figure 39. A shell was about to be deposited on an egg in the oviduct. Other eggs in various stages were in the process of development. Several ovulated follicles were visible, and the largest one (indicated by an arrow), probably was the source of the egg about to be laid. This mallard hen had laid four eggs when she was killed. Since I wished to compare the condition found in this laying mallard hen with a hen that I believed to be nearly through laying her clutch of eggs, I collected a gadwall hen on June 19, 1950, as she settled down on a nest at 7:00 a.m. The nest held 10 cold eggs when she arrived. Since the clutch size for Delta gadwalls did not exceed 11 eggs, I believed that the hen was ready to lay her last egg.

When I autopsied this hen, I found a completely developed egg ready to be laid but found no follicles over 5 millimeters in diameter. Incubating hens were collected also. It was found that a rapid regression in the size of follicles occurred when incubation began.

From this anatomical evidence, it seems likely that continuous laying cannot occur when the nest is destroyed at the time that the last egg is about to be laid, and definite that it cannot occur when the nest is destroyed during the incubation period. Just when, in the laying period, clutch size is determined is unknown. But it is evident, that regression of follicles and their increase again to egg-laying size happen rapidly.

The lengthening period of the renesting interval as the ovaries regress during the incubation period is consistent with anatomical data existing on domestic fowl. Aside from the fact that the smaller ova have more growing to do than the larger ones, the rate of growth of the larger ova has been found to be faster. Of this, Riddle (1916: 388) says, ". . . the rate of growth is not uniform, or continuous, and may be divided into a number of periods. In a previous study, I found some evidence that the growth period in the fowl is not a steady, unbroken one—not even among the larger continuously growing ova of a functioning ovary; but at a certain point, the rate changes quite suddenly to a rate apparently eight to 20 times faster than before; and further, that the change in rate normally occurs five to eight days before ovulation, and when the ova have reached a diameter of about 6mm."

Renesting after loss of brood.—In 1949 and in 1950, I made an effort to determine what happened to a hen after her brood was taken from her.

In order to answer this question, I nest-trapped and marked 10 hens and left them to incubate their eggs. Then, just as the young had hatched and were dry enough to leave the nest, they were taken from the hen. The procedure was carried out with six pintails, three blue-winged teal and one mallard. Of this group, two of the pintail hens were known to renest following loss of brood. One waited for a period of 18 days and then laid a clutch of three. Her first clutch had been seven. The second waited for 16 days and produced a clutch of eight. Her first clutch had been 10.

From these observations we see that some hens will renest even after their eggs have hatched and the brood is destroyed. How long after hatching a hen will do so is unknown. Variation in the reaction of the hen to early-season brood-loss and late-season brood-loss may be considerable. This likewise is unknown.

Location of renests.—The locations of first nests and renests of 31 hens are plotted in Figure 40. Of these hens, all but one had located their second nests within a relatively short distance from their first nests. The maximum, minimum, and average distances between first nests and renests of 31 hens of five species are given in Table 22. Only one hen moved over 700 yards. This bird was a pintail which located her second nest 1,500 yards from her first nest. In this case, rising marsh waters flooded the area around the site of the first nest so that the hen was forced to move.

It is of interest to note that the pattern of nest sites in Figure 40 (first nests and renests in a single year) is similar to the pattern of nest sites in Figure 7 (nests of returning hens, two or more years).

Number of unsuccessful hens which renest.—How many unsuccessful hens renest is unknown. The fact that only 42 of our 220 marked hens were known to do so does not mean that the others did not. I have no doubt that there were renests on the study area that I did not discover. This is indicated clearly by the fact that six of my 42 renest records resulted from seeing a brood with a marked hen rather than by finding the second nests. Other hens may have nested outside the study area; still others may not have renested at all.

TABLE 22. DISTANCES IN YARDS BETWEEN DIFFERENT NESTS OF THE SAME HEN DURING THE SAME YEAR

Species	Number of Hens	Maximum Distance	Minimum Distance	Average Distance
Mallard	1	700	700	700
Gadwall	3	600	45	320
Pintail	15	1,500	85	282
Shoveller	7	765	135	355
B-w Teal	5	405	135	270

Low (1945:48) in his study of the redhead divided the nesting season into 10-day periods and showed the number of hens in each period which terminated their breeding efforts because of nest failures.

Laven (1940:131) in his study of the ringed plover in Germany says that one pair produced four additional layings after the loss of its first clutch, while other pairs in the area produced one clutch and departed when that was lost. He also points out that great variation between individuals and birds of different ages must be expected.

That some pairs of ducks wander following nest destruction may account for the large gatherings of pairs which one often sees on the breeding grounds late in the nesting period. As has been mentioned, this may follow changes in environment of first nesting. Arthur S. Hawkins and I saw such a gathering of pairs and lone hens of gadwalls, pintails, mallards and blue-winged teal at Whitewater Lake, Manitoba, on June 26, 1946. For three springs, I saw such gatherings of pintails south of the study area at Delta. Similar flocks have been observed by Stoudt and Davis (1948:134) who say, "Apparently the South Dakota area had a large number of either transient or non-breeding birds on the transects during the first week in June and these consisted mainly of pintails, mallards and shovellers . . . It may be that this state has more than her share of non-breeding birds which may consist of either infertile or sexually immature ducks or bachelor drakes. . . . The fact that 135 pairs of pintails were seen on Transects F and B in June and only four broods were found in July illustrates the need for future studies along these lines."

Renesting and inventory counts.—The presence of birds of undetermined status as described above may confuse the counting of breeding pairs for inventory purposes. Smith and Hawkins (1948:62) ask the question, "How many times during the season does the population of an area renew itself; *i.e.*, what is its turnover rate? If a given drake defends his territory against all other drakes of his kind for only two weeks and the nesting period for the species spans six weeks, there is the theoretical possibility that the area appraised at ten pairs per section had a total season's population of thirty pairs (three turnovers)."

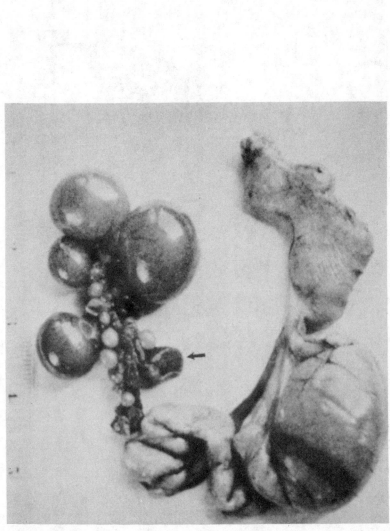

Figure 39. Reproductive tract of a laying mallard hen. The arrow indicates an ovulated follicle from which the large egg in the oviduct probably came.

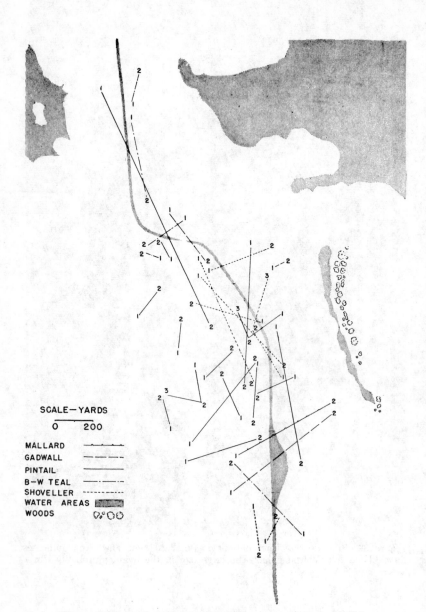

SCALE—YARDS

0   200

MALLARD   —·—·—
GADWALL   —   —   —
PINTAIL   ————
B-W TEAL   —·—·—
SHOVELLER   -------
WATER AREAS   ▓▓▓▓
WOODS   °°°°

Figure 40. Locations of first nests and renests of 31 hens of 5
species, 1947-1950.

To what extent this wide span of nesting is accounted for by late first nesters, and to what extent by renesters who failed earlier, is not known. If one considers the ducks' persistence to renest, however, it seems likely that a large number of the late nesters are merely the same birds which failed earlier. Hence, rather than having a "turnover" of hens, we have an extension of the nesting season brought about through renesting.

Persistence in renesting.—Not all species exhibit the same persistency to renest following the destruction of their first nests. The measurement of this comparative persistency is not made easily, however, because of several factors. The number of hens marked in the renesting study was not the same for all species. The time of season in which they were marked was not always the same. Many hens probably were marked for the first time when they already were on their second nests. Mallards were difficult to handle, deserted readily, and did not offer good opportunity for comparison.

In my studies, I repeatedly found that of all species, the pintail was the most persistent renester, and the blue-winged teal the least persistent renester. Of 62 pintail hens marked, I obtained definite renesting records on 19, or 30 per cent. Two of these were known to renest a second time following destruction of their second nest. It was the pintail which gave me positive information that renesting not only follows nest destruction but sometimes also follows destruction of a hatched brood.

In the five years of my study, I nest-trapped and marked 88 blue-winged teal hens. Of this number, only five, or about 6 per cent, were known to renest. This sample of marked hens was larger than the number of marked hen pintails, but the renesting number was only about one-sixth as great.

Only one renesting mallard was recorded of 20 hens marked. Of 33 shovellers marked, only seven were known to renest. One of these, however, was persistent enough to renest twice in one year. This shoveller, and two pintail hens, were the only three individuals in my study known to do this. The gadwall was a persistent renester; 23 per cent of 16 gadwalls marked were known to renest. For these three species just mentioned, a relatively small amount of data was obtained. Although only one mallard was known to renest, I am certain, from dates of nesting, that they readily do so. Williams and Marshall (1938:41) working at Bear River, Utah, found that 70 per cent of the duck eggs found hatched successfully. Sowls (1948:130), however, working at Delta, found that only 35 per cent of nests found hatched successfully. Thus we see that there is a variation in early-season nest losses between regions and between years.

In 1940, heavy early-season nest losses at Delta resulted from spring floods. In 1946, spring fires wiped out all early nests over large parts of the Delta marsh, and only renesting could have maintained population numbers that year. In 1950, spring plowing on the Portage plains took an abnormally heavy toll of pintail nests during the peak of the nesting season.

Evidence that early losses are compensated for by renesting is obtained from an analysis of nesting data. To interpret such data we might set an arbitrary date after which all nests started would be considered renests. For the Delta marsh area, I would set this date for mallards and pintails at May 20. (This date would vary with early and late years but would be correct for the latest season I have seen at Delta.) From Figure 21, it is evident that 17 per cent of all mallard nests, and 25 per cent of all pintail nests in 1949, were started after this date and would therefore be considered renests. In 1950, 48 per cent of the mallard and 44 per cent of the pintail nests were begun after May 20 and would be, therefore, renests. I believe that this is a conservative interpretation. As has been pointed out earlier, these figures do not include renests of marked hens which renested after my disturbance of the first nest. Furthermore, nests destroyed very early in incubation and nests destroyed during the laying period could have led to renestings which would have been begun before this date.

For the later-nesting blue-winged teal, I would set June 24 as the arbitrary date after which all nests begun could be considered renests. Figure 21 then shows us that 9 per cent of the blue-winged teal nests found in 1949 were renests, and in 1950, more than 30 per cent were renests.

I believe that the data on the gadwall and shoveller given in Figure 22 are insufficient for such an analysis.

During the spring of 1950, I determined from embryo dating of eggs found in early-season nests that a large hatch of mallards and pintails would occur during the period between June 6 and June 12. This prediction was fulfilled when the majority of early-collected eggs hatched in the Delta incubator during this period. Contrary to the prediction that large numbers of pintail and mallard broods would appear in the wild, no broods were seen until July 8. At least a half-dozen observers had been in the field during this time. I believe that only on the basis of early and heavy nest loss can this lack of early broods be explained.

Analysis of brood data on the dates of hatching within arbitrarily set periods would give an estimate of the percentage of young which resulted from renesting. In my study, I was unable to gather enough brood data to do this.

Although we cannot determine from the data now available the percentage of thwarted hens which renest, and the percentage of production which results from renesting, it is clear that renesting at Delta is of major importance in maintaining the waterfowl population there.

SUMMARY

1. Ducks have but one brood a year but readily renest if their first nest is destroyed. While the extent of nest destruction has been studied thoroughly in the United States and Canada, the details of the renesting phenomenon have received slight attention.

2. In 1946, a study to answer basic questions in regard to renesting in ducks was started at Delta, Manitoba. Hens were captured on their nests, banded, feather-marked, their nests robbed, and hens released. After this operation, the area was watched for renesting birds.

3. Of 220 birds so trapped, 42 were known to nest again. Of the 42 renesters, three hens were known to have three nests during one season. Two of these were pintails and one was a shoveller.

4. Although the average number of eggs in first clutches is greater statistically than the number of eggs in renests, the difference is not great enough to distinguish first clutches from renests. Some renests have as many or more eggs than other first nests.

5. Renests cannot be distinguished safely from first nests by their appearance. Many renests look like first nests.

6. The time between the destruction of the first nest and the

laying of the first egg in the second nest (*the renesting interval*) varies with the stage of incubation at which the nest is destroyed. The farther advanced incubation is, the longer the hen waits before renesting. All hens whose nests had been destroyed after incubation had begun, waited at least three days before laying again. For each additional day of incubation hens waited an average 0.62 days before renesting.

7. Continuous laying without a renesting interval occurs when the first nest is destroyed during the early part of the laying period. A renesting interval probably is required before laying again if nest destruction occurs on the day the last egg of the first clutch is due to be laid.

8. Two pintails, whose young were taken from them after hatching, renested following the robbing of their young. Thus, renesting may follow the destruction of a hatched brood.

9. The percentage of unsuccessful hens which renest has not been determined. In this study some may have renested outside the study area and some may not have renested at all.

10. Hens located their renests in the vicinity of their first nests. All hens located their renests within 700 yards of their first nests except one pintail which moved 1,500 yards to renest.

11. Of the five species studied, the pintail was the most persistent renester, and the blue-winged teal the least persistent.

12. Inventory counts made by the transect method are complicated by the renesting phenomenon in that birds counted as breeders at one time may be counted as breeders again at a later time when they are in reality renesters that failed earlier.

13. Renesting varies in its importance to production with locality and year depending upon the extent of nest loss. In some areas during some years it may account for nearly the entire production.

14. The number of nests found which were renests was estimated by regarding all nests begun after an arbitrarily set date as renests. The arbitrary date for mallards and pintails at Delta is May 20; for blue-winged teal, June 24. Thus, in 1949, it was estimated that at least 17 per cent of all mallard nests, 14 per cent of all pintail nests, and 9 per cent of all blue-winged teal nests were renests. In 1950, at least 48 per cent of the mallard nests, 44 per cent of the pintail nests, and 30 per cent of the blue-winged teal nests were estimated to be renests.

CHAPTER X

HEN AND BROOD BEHAVIOR

*T*HE BEHAVIOR of the hen and her young during the brood season lacks the aggressive aspects of the hen's and drake's behavior during the territorial period. It lacks the complete solitude of the incubating period but does not approach the gregariousness which is typical of all waterfowl in autumn.

Broods are raised in waters bordering or near to the meadows where they hatched. There they must survive many dangers between hatching and flying. They survive only because of their instinctive and learned behaviors associated with the ability to escape danger, to hide, and to follow the protective calls and signals of their mother. From the simplest instinctive response to the complex learned behaviors which develop slowly with time, all play an important part in survival. The climax of this period comes when the instinct to fly develops by practice into strong, alert and precise flight.

Hatching.—In the incubator at Delta, thousands of clutches of wild duck eggs have been hatched; and it has been easy to watch and record the hatching process. In the wild, however, and under natural conditions, few hatching nests have been observed. In my study at Delta, I watched four pintail nests and two blue-winged teal nests from the time the first egg pipped until the young were ready to leave.

In 1949, I recorded the incubator hatching of 15 clutches. Of this group, seven clutches had a hatching period of less than

143

one day, all eggs of each clutch hatching on the same day.
Two clutches required a period of two days and six required a
period of three days. Of six clutches hatching in the wild, all
eggs of each clutch began pipping at the same time, and all
young emerged within an hour of each other. Clutches hatched
in the wild and under natural conditions hatched uniformly.
Some incubator hatched clutches required up to three days to
complete the process.

Brood movements.—The intentional robbing of nests in the
renesting study at Delta curtailed the collection of mass brood-
movement data. On the study area, I saw one pintail hen
move her brood 800 yards within the first 24 hours after
hatching. Several others moved their broods lesser distances to
the nearest slough or ditch. Three pintail broods that were
hatched by marked hens on the study area were known to
grow to flying age within 500 yards of the nest site.

Important information on brood movements has been given
by Evans (1951) who studied the movements of marked
ducklings in the glaciated pothole country near Minnedosa,
Manitoba. Evans found that movements between potholes
occurred in a random direction. Of pintail broods he wrote
(pp. 47-48): "This brood travelled 0.42 miles from the nest in
16 days. . . . These three broods were observed to travel 0.78
miles in 71 days for an average of 0.011 miles per day. No
brood of this species was known to occupy a single pothole for
more than 14 days." Of mallard broods, Evans (p. 49) said:
"These two broods travelled a total of 0.48 miles in 38 days
for an average of 0.013 miles per day. . . . No mallard brood
was known to occupy a single pothole for more than 20 days."
And Evans (p. 49) wrote about the blue-winged teal: "The
longest time a brood of this species was known to occupy a
single pothole was 28 days." Evans (p. 50-51) also observed a
shoveller brood that spent 16 days in one pothole, and a gadwall brood that spent 17 days in one pothole. He was able to
determine which species moved their broods most readily.
"Pintail broods were the most mobile, followed in order by
canvasback, mallard, redhead, blue-winged teal, and baldpate,
while ruddy ducks broods were the least mobile," according to
Evans (p. 108).

Brood reactions to calls of hen.—The reactions of the young
to different calls of the hen were observed in the field. There

was the freezing reaction in which the young remained quiet and immobile, and the huddling reaction in which the young moved close to the hen and stayed near her in a solid group.

On July 13, 1950, I attempted to identify a pintail hen which was known to be banded and which had a brood with her. She was known to be somewhere in a small whitetop-edged pothole with nine ducklings about three weeks old. At 6:45 a.m. two other observers and I stationed ourselves at three points around the pothole and slowly drove the hen and brood out into the center where there was open water. The hen and brood had been hidden in heavy flooded whitetop on the south side of the pothole. The hen flushed and issued a long rasping call, which apparently served to alert the young; all of them remained stationary with heads high. They made no effort to join the hen or to escape to cover. After several minutes we withdrew to about 40 yards from the pond and allowed the hen to return. After returning to the pond, the hen continued to emit her loud, rasping call and the young remained immobile and alert. The hen moved slowly to the south side of the pond and changed her loud rasping call to one resembling a low *cheep-cheep*. At that moment, the entire brood dashed toward the flooded whitetop cover.

When working in heavy stands of flooded whitetop on the study area, I often heard the cheeping of ducklings. But as soon as a disturbed mother took to the air and circled the area making loud rasping calls, the young ceased their cheeping and were quiet.

The reactions of the young to the calls of the hen apparently were learned during the first few hours after hatching. This point has been demonstrated adequately by various students of bird behavior. Lorenz (1937:262) believed: ". . . that most birds do not recognize their own species 'instinctively,' but that by far the greater part of their reactions, whose normal object is represented by a fellow-member of the species, must be conditioned to this object during the individual life of the bird." In his experiments with young geese and ducks, Lorenz found that in some cases he was able to induce young birds to take for their parent-companion whatever living thing they were exposed to first. The process whereby this attachment originated, he called "imprinting."

Nice (1953:33) has called the early hours of life a period of rapid learning of the characters of the parent and credits Heinroth as the first to discover that the newly hatched grey goose *(Anser anser)* adopts as its parent the first living being it sets eyes upon. Tinbergen (1951:150) has termed these first few hours of an animal's life a period of critical learning.

Fabricius (1951) was able to imprint young ducklings to follow him and come to his calls, and also to get young ducklings of one species to be imprinted to the older ducklings of another species. He found that the period during which young tufted ducks could be imprinted was from birth up to at least 36-38 hours. Both Fabricius and Lorenz focused considerable attention on the releasing mechanisms of the "following" reactions. Nice (1953), using the techniques developed by Fabricius, succeeded in imprinting to human beings 12 ducklings of five different species. These ducklings accepted human beings as parent-companions when subjected to visual signs and acoustic signals during the first few hours of life.

During my studies, most nests were robbed intentionally in order to gather data on renesting. Consequently, few nests were left to be observed as they hatched normally. In wild nests, I found that the young showed no fear reactions immediately upon hatching and while still wet. After the young were dry and presented a downy appearance, they showed an awareness of intrusion and responded by attempting to scurry out of the nest to hide in the grass at the edge of the nest.

Figure 41. Brood of blue-winged teal ducklings one hour after hatching.

Figure 42. Same brood of blue-winged teal eight hours after hatching.

Figures 41 and 42 show a brood of blue-winged teal young just after hatching, and the same brood eight hours after hatching. Although some variation existed in the time required for drying off and leaving the nest, I suspect that most ducklings left the nest within 12 hours after hatching. The drying-off period normally was spent cuddled under the hen.

Tolling of intruder by hen.—A hen is said to toll when she diverts an intruder's attention from her brood by moving deliberately and conspicuously from it. While working a dog in the nesting meadows at Delta, I saw on many occasions hens try to draw my dog away from a brood that was hidden in the grass. The most spectacular observation of this occurred on July 5, 1948. On that afternoon, I saw a pintail hen circling and quacking loudly over my head as I walked through a flooded whitetop meadow on the study area. I could identify the hen by the colored bands on her legs and was anxious to learn whether she had renested and now had a brood in the heavy grass. When I encouraged my dog to find the young, the hen immediately began luring him away. She accomplished this by swooping low over the dog and flying slowly in front of him in a direction away from where the brood apparently was hidden. By continued search one young pintail was found.

The broods of young ducklings show a remarkable timing or synchronization of activities. In his descriptions of the "imprinting" of young ducklings, Fabricius (1951:165) has made note of this: "All the young birds of one group generally perform the same kind of activity simultaneously, such as preening, eating or sleeping. This synchronization is very important for successful conduction of a brood." Just as it is important to the investigator who has conditioned young birds to follow him, so it is important to the safe conduct of the brood when led by their mother in the wild.

Hiding of young by hen.—On a number of occasions it was clear that the mother intentionally hid her young before attempting to toll away an intruder. Sometimes the hen alerted them with a call producing the freezing response. Sometimes the hen actually guided the young into a safe place before attempting to lead the intruder away.

On a July day in 1950, Dr. W. J. Breckenridge, James

Houston and I watched a redhead hen and brood of eight
4-day-old ducklings. The hen was swimming down the middle
of a ditch with the young close by her. We watched her for a
short time until she swam into a small side channel where the
ditch had overflowed and flooded phragmites cover was available. Almost immediately the hen swam out alone and continued down the ditch while the young remained hidden in dense
reeds.

Hiding of hen with brood.—In my observations at Delta, I
found that a hen, when disturbed in a flooded area, usually
attempted to hide her young and then draw me away from the
place where the young were hidden. Sometimes, however, the
hen hid herself with her young.

On June 7, 1947, a hen mallard with a brood of 8-week-
old ducklings was seen in a part of the Delta ditch where the
edges were relatively bare because of grazing. This mallard
hen led her brood down the ditch for 20 yards and then
attempted to hide them against a tiny clump of wild barley on
the bank. In doing so, the hen stretched her neck out over the
shore in the shadow of the grass clump and remained quiet.
The young huddled close to her.

When puddle-ducks with young were disturbed by a canoe,
hiding behavior was more common than tolling, or leading
away. But when they were disturbed in shallow, flooded
meadows, tolling was to be expected. When diving-ducks with
young are disturbed, they take their young to open water
instead of into reeds (Hochbaum 1944:105).

Feigning behavior of hen.—A hen is said to feign when
she exhibits a spectacular flapping movement across water or
land similar to the escape behavior of flightless molting birds.
This behavior serves to attract attention away from a hen's
brood as in tolling, but the hen uses a flapping movement
across land or water, whereas in tolling she swims or flies.

Hochbaum (p. 105) describes the feigning of puddle ducks
and says: "Feigning behavior is most intense in pintail and
blue-winged teal . . . and is less intense in the gadwall, shoveller and mallard." I have seen feigning behavior in all species
at Delta on open water and on dry land but have not witnessed it in flooded meadows where the hen invariably took
to the air and attempted to lead me away by flying.

The feigning behavior of a hen with a brood is the same as that which frequently is exhibited by incubating hens during the late stages of incubation.

Defense of young by mother.—Hens have been known to defend their young against intruders without making an attempt to draw the intruder away. The best example of this that I know of was an instance in which a mallard hen successfully defended her brood against a mink. The observation was made at Delta and was described to me in a letter dated January 22, 1947, by Arthur S. Hawkins who wrote:

"On June 25, 1946, I saw a mink attack 8 newly hatched ducklings belonging to a mallard. The fight which lasted at least 10 minutes was staged within 50 yards of my observation station and at times it came much closer to me.

"The hen's strategy was to keep her brood bunched. Each time the mink approached she swam between her brood and the invader. She quacked constantly and flapped her wings repeatedly in his face giving the ducklings a chance for a wild dash across the ditch."

Brood habitat.—Areas of flooded vegetation formed the most important brood cover at Delta. On the study area, large flooded fields of whitetop were preferred rearing grounds for the surface-feeding species. Open waters were not used extensively by the surface-feeders with broods, but they were used by diving duck broods. In the flooded grassy meadows, small hummocks formed loafing places where the mother and young spent much of their time. These dry loafing places are important to the young ducks and are a necessary part of any rearing ground. In the larger bays, muskrat houses and mud bars served this purpose. On the long narrow ditch which ran through the study area, the loafing logs and platforms put out for breeding-season adults were used extensively by the broods as loafing sites.

In the pothole country of Manitoba, Evans (1951:104) found that the large open potholes with sedge-whitetop margins were favored as brood rearing areas while similar areas with bulrush cover were a close second. He also found that depth was an important factor influencing pothole selection; and that during periods of low water, potholes over two feet deep definitely were preferred to shallower areas.

Summary

1. The hatching of clutches in the wild and hatching in the Delta incubator were compared. Six clutches hatched in the wild and under natural conditions hatched uniformly. Some incubator-hatched clutches required up to three days to complete the process.

2. The first 12 hours of the life of a newly hatched brood was observed in the wild and the behavior noted.

3. The reactions of the young to the calls of the hen apparently were learned during the early hours of life.

4. Brood movements were watched by me at Delta and by Evans at Minnedosa. Broods moved in random directions and varying distances. One pintail hen moved her brood 800 yards within 24 hours of hatching. Three pintail broods grew to flying age within 500 yards of the nest site. Evans found pintail broods to be the most mobile, followed by canvasback, mallard, redhead, blue-winged teal, baldpate, and ruddy duck.

5. Brood reactions to the calls of the hen were watched in the wild. There was a "freezing reaction," "huddling reaction," and a reaction in which the young joined the hen and swam with her. The reactions of the young to the calls of the hen apparently were learned during the first few hours after hatching.

6. Sometimes brood hens "toll" when intruded upon. By moving deliberately and conspicuously from the intruder they attempt to divert attention from their broods.

7. Hens were known to hide their broods before attempting to toll an intruder away. Sometimes they hid themselves with their young.

8. Among the surface-feeders, hiding behavior was to be expected when water was deep; tolling behavior was to be expected when it was shallow.

9. Feigning behavior differs from tolling behavior in that a feigning hen exhibits spectacular flapping movements across the water or land, whereas a tolling hen swims or flies.

10. Hens sometimes defend their young against intruders.

11. Areas of flooded vegetation, such as whitetop, were preferred rearing grounds for the surface-feeding species at Delta. In the pothole country, large open potholes with sedge-whitetop margins and good depth (over 2 feet) were preferred.

CHAPTER XI

AUTUMN BEHAVIOR AND THE SHOOTING SEASON

*A*S SUMMER passes on the northern breeding grounds, landscapes change. The dry, brown nesting cover of spring is replaced by waving green grasses and a great variety of other plants. Shrunken to half their size, ponds expose their muddy banks; and pondweed, milfoil and bladderwort choke their centers.

The habits of ducks also change. Pairs no longer claim loafing edges; the business of breeding is over; and the air is empty over the big meadows. But in the bays and sloughs, thousands of ducks gather in great flocks. Gregariousness is the rule for waterfowl throughout most of their lives. With the breakdown of territories and freedom from breeding-season home ranges, ducks seldom are seen in small groups. They are in flocks for the remainder of the summer on the northern marshes, through the fall-flight south, during the winter and northward flight until another breeding season.

Signs of autumn behavior.—Almost before spring had fully arrived, the first signs of summer and fall behavior appeared. In May, the first drakes to complete their breeding activity began to pass into their post-breeding phase; and they were seen together in twos and threes along pothole rims, water-filled ditches and on slough banks. Here they often took part in the pursuit of teasing hens, and some of them mated with renesting females. The drakes joined others in small groups and eventually made up the flocks of thousands which moved into large marshes for the flightless period.

Those that did not participate in late-season courtship

probably had passed through the breeding period and no longer were in breeding condition.

In my studies of behavior, plumage was an aid to interpreting the actions of ducks during the later part of the nesting season. The passing of the nuptial plumage and the coming of the eclipse plumage corresponded, roughly, to the end of breeding for individuals. It was, however, impossible to correlate precisely breeding condition and plumage change; and I was unable to tell from the appearance of a live bird what breeding condition he was in. The breeding plumage was reached in the fall or winter, long before the gonads had enlarged to the size they attained during the breeding season, and while the reproductive glands still were non-functional. In accordance with the conclusions of Holm (1947), who found that the period of spermatogenesis in the mallard extended into the period of eclipse plumage, I observed at Delta that some drakes of shabby plumage remained with their hens while the hens produced fertile eggs.

That there is a more rapid change to the eclipse plumage by breeding birds than by the non-breeders has been pointed out by Hochbaum (1944:112) who wrote: "In our captive ducks, the first to enter the eclipse are the earliest mated drakes. Drakes which breed late follow, and the last to assume the eclipse are non-breeding males." I observed this to be true of wild lesser scaup drakes. Mated drakes were the first to acquire eclipse plumage in spring, and those that did not have mates were later; although, from collections of both, it was determined that this latter group had greatly enlarged testes and apparently were in breeding condition.

The gathering of "thwarted" pairs.—The gathering of ducks near the molting grounds was not restricted to groups of lone birds, although this was the general rule. There was considerable evidence, though not clearly understood, that many ducks came into the molting area as pairs.

On June 26, 1946, at Whitewater Lake, Manitoba, Arthur S. Hawkins and I estimated that there were 125,000 ducks on that famous marsh. We had expected to find a concentration of drakes, but found lone drakes and pairs. It seemed likely that the mated birds had ceased nesting attempts while still paired and had moved to the molting area together.

Similar gatherings on a smaller scale were seen each year at Delta. These groups included unpaired drakes, lone hens, and pairs. During the summer of 1950, pintails began gathering on a flooded pasture on the Delta study area during the second week of June. Data on the number of hens seen there are given in Table 23. These figures indicate that, for the sample of pintails counted in the Delta marsh at the close of the nesting season, approximately one quarter were hens which had either not nested or had ceased trying.

Duration of the molt.—The first of the puddle ducks to molt were ready to drop their primaries and grow new ones in June. Hochbaum (1944:121) says that some captive mallards, pintails and blue-winged teal completely renew the primaries within two and a half weeks after the flight feathers are dropped, but that he believes that between three and four weeks is the usual period.

In 1950, I kept seven wild pintail drakes in captivity in order to measure the flightless period. There was no great difference in the rates of development of new primaries of any of these birds. Five of them flew out of an open-topped pen on the twenty-seventh day, while two did not fly until the twenty-ninth day. Thus, for pintail drakes, the flightless period was close to four weeks.

Last to molt were the late successful hens. In the extensive

TABLE 23. SAMPLE COUNTS OF THE SEX-RATIO OF PINTAILS IN POST-BREEDING SEASON GATHERINGS AT DELTA, 1950

Date	Drakes	Hens	Pairs	Percentage of Hens
June 13	31	3	21	31
14	52	3	16	21
15	42	0	18	23
16	50	3	23	26
19	55	5	15	22
20	76	9	17	21
23	86	16	18	24
Totals	392	39	128	Ave: 24

hunters' bag check made at Delta only one surface-feeding duck still in the "flapper" stage was found. This bird was a gadwall hen having new blood quills about 2 inches long. She was found in 1946 on October 10.

Location of molting area.—We knew that the number of birds molting in the Manitoba marshes exceeded the number which nested there. The number of birds appearing in these marshes in July and August was far greater than the resident population. Too, some species, practically absent as breeders in May, appeared in large numbers at Delta in July. This was true particularly of baldpates and green-winged teal. Thousands of these ducks must come long distances to molt and probably come from many places.

That this is true has been pointed out by Hochbaum (1944:116) for Manitoba, and Stresemann (1940) for Europe and Asia. The latter writer describes the concentrations of birds in the Volga and in Tibet, which gather there, far south of their breeding marshes, to molt.

Not all ducks move long distances to complete their summer molt, however; some adult hens that nest in large marshes obviously molt their primaries near the place where they nest. This is evident in the return of bands from birds shot in the Delta marsh. Two hundred and twenty adult hens were trapped and banded on the study area. From this group, hunters submitted 12 band returns, three of which were shot in the Delta marsh; the hens apparently spent their flightless period there after nesting. One of these was a shoveller banded on the study area on May 22, 1949, recaptured there in 1950, and finally killed there on October 18, 1950, not over 400 yards from her nest sites of 1949 and 1950.

These records indicate that some adult nesting hens spend their flightless period in the same marshes where they nest, and sometimes quite near their nest sites.

Population build-up in late summer.—The population build-up at Delta was studied by the Station staff and by the U. S. Fish and Wildlife Service flyway biologists during the period of this study. Hawkins and Cooch (1948:98) showed the 1947 and 1948 Delta Marsh populations. These data, plus some later season counts which they did not give, are shown in Figure 43 to illustrate the population's rise and fall. The peak at Delta was accounted for mainly by mallards and pintails. By early October, many mallards and most of the pintails were gone.

Feeding flights to grain fields.—Occasionally in April and May, before the break-down of spring gregariousness, and commonly from August until freeze-up, after the resumption of gregariousness, daily stubble flights of mallards and pintails occurred.

The habit of going to the grain fields to feed undoubtedly is one that had been learned over a period of years. That this habit began several years after the prairie country was settled, and long after grain first was raised there, is evidenced by the observations of hunters who remember these marshes as they were 50 or more years ago. James A. Munro (verbal communication of January 25, 1952) recalled that when he first shot ducks in Manitoba in 1903, mallards were hunted exclusively in the sloughs and marshes and did not at that time make flights to "stubble" to feed, although barley, wheat and oats were common crops then on the neighboring Portage Plains.

The same story was told to me by Frank A. Farley of mallards near Camrose, Alberta. Of the pintails in that region, Farley (1932:20) said, "Pintails are slowly learning to feed in the grain fields, in the same manner as the mallards."

G. Gillis Tidsbury, of Portage la Prairie,, Manitoba (letter of February 1, 1954), says "I would say that mallards and pintails have been feeding on stubble in this area almost from the time that the land was first cropped in the area that was adjacent or within a few miles of the marsh, but it is only in what you might call recent years that the gun pressure has become so intense that they have started going out ten to twenty to thirty miles from the marsh."

Though some disagreement seems to exist between early observers, it is clear that the flights to grain fields are more conspicuous now than they were in the days of early agriculture in the prairie provinces.

Both mallards and pintails were stubbling species at Delta. As Hochbaum (1944:134) has pointed out, there are two groups of mallards in the Delta region in late summer and fall; those that go to the grain fields regularly to feed and those that do not. Some units of the population definitely do all of their feeding in the marsh and never make flights

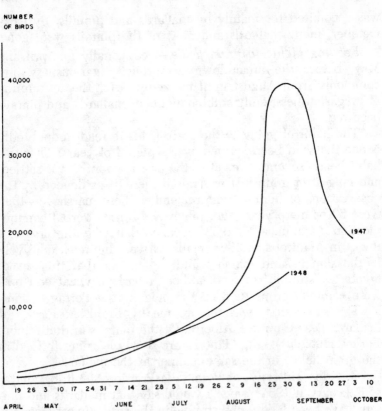

NUMBER
OF BIRDS

40,000

30,000

20,000   1947

1948

10,000

19 26 3 10 17 24 31 7 14 21 28 5 12 19 26 2 9 16 23 30 6 13 20 27 3 10

APRIL   MAY   JUNE   JULY   AUGUST   SEPTEMBER   OCTOBER

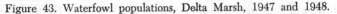

Figure 43. Waterfowl populations, Delta Marsh, 1947 and 1948.

to grain stubble. Other large units of the fall population make daily flights from loafing sandbars to the neighboring grain fields.

When the flights to stubble began in August, the mallard-pintail ratio was nearly even; but, as the season progressed, the ratio began to shift in favor of the mallards. After early October, pintails, some of which departed early from the northern marshes, were seen only occasionally among the large number of birds flying back and forth between loafing sandbars and grain fields.

Although some changes in daily routine probably occurred because of weather, hunting disturbance, and other factors, there usually were two flights to the grain fields each day. The first of these flights took place in the morning while the sky still was dark. Some of the birds were back on the loafing bars before the sun was up, but most of them came in to the loafing place shortly after sunrise. In the fall of 1949, I set up a blind on a heavily grazed mudbar which accommodated about 3,000 ducks. On August 31 of that year, I entered the blind at 3:45 a. m. and found no birds on the bar. At about 15 minutes to sunrise (sunrise 5:12 a. m.), small groups of mallards and pintails began to arrive. By 8:00 a. m., thousands of ducks were there, and no more birds came in. This pattern of early morning feeding held true for this particular group of birds for the three weeks that I kept them under observation; and I believe that it is fairly representative of the pattern of early-morning feeding. I have, however, seen stubbling mallards and pintails on the grain fields at almost every hour of the day, indicating that there is variation from that pattern.

The evening flights are known better to hunters and residents on the prairies. Most mallards and pintails left the loafing bars in the afternoon between the hours of 4:00 p. m. and 8:00 p. m. During that period, long strings of ducks could be seen coming and going along one line of flight. At Delta, the beginning of the flight, as it left the marsh, followed predictable flight lanes. These conformed to the long winding creek beds below, which extended from the marsh into the prairie. The flight lanes were used conspicuously in the days prior to the opening of the hunting season. After gunning began, considerable dispersal occurred and the flight lanes were adhered to less conspicuously.

In the fall of 1949, I observed regular stubble flights, predictable as to time and place, until September 20; but after that date, there was a widespread diffusion of flocks and a disruption of flight lanes. This occurred 10 days before the opening of the hunting season on the prairie that year and could not, therefore, be attributed to gunning. In this instance, the important factor contributing to the change seemingly was the plowing of much of the prairie and the shifting of suitable feeding areas.

As to the number of birds using a loafing area, and the number in the flocks that make stubble flights, there was con-

siderable variation. The flock that I watched in 1949, on the mudbar where I had constructed a blind, contained about 3,000 birds. Another group, that moved two miles farther north on the lake shore, numbered about 6,000. This latter group was counted on several occasions. On the evening of September 6, a total of 6,194 ducks was counted returning from the fields to their loafing bar on the lake shore between 7:15 p. m. and 8:00 p. m., when the flight ceased. On the following evening a flight began at 4:55 p. m. and 3,278 mallards and pintails were counted going to the stubble up until 7:35 p. m. The flight back began at 7:15 p. m., and 5,887 mallards and pintails were observed returning up to 8:15 p. m. None were seen returning to the marsh after this time, although a watch was kept until 8:30, when it became too dark to see the birds. In some marshes, the stubbling flights may be much larger.

As one watches mallards and pintails returning from the fields, it is not uncommon to see some so full of grain that the swollen crop is conspicuous. This is a common sight known to most hunters and wheat farmers on the Portage Plains.

From my blind on the loafing bars, I observed that the birds usually alighted on the water, drank, then slowly made their way to the edge and climbed up onto the bare ground. There they rested, some sleeping, sitting so close together that they almost touched each other. No aggressive behavior was exhibited.

The loafing places, where the grain-feeding ducks spent most of their time during the late summer and early fall, were not always the same from one year to the next. The southern shores of Lake Manitoba offered many attractive sandbars in the years 1946 and 1947, and they were used heavily by ducks during the summer season. But in 1948, higher lake levels annihilated these places, and the loafing areas for the years 1948 and 1949 were located on the edges of large bays in the Delta marsh where grazing and watering cattle had created open mud-flats.

Since resting places are necessary to waterfowl, both during the nesting season and during the late summer and fall, they must be considered in any management plan. Deep water with densely vegetated shores will not attract and hold high populations of surface-feeding ducks during late summer and fall. Where the natural lowering of water levels occurs, suitable shorelines for resting ducks usually are present. On many areas of stable, deep water and dense vegetation, open loafing grounds can be provided easily by using limited numbers of livestock to graze shorelines. The cattle are effective in destroying undesirable heavy growth and producing the desired bare beaches for resting ducks. As surely as the buffalo in pristine times made its impression on the marshes; today's cattle have their place in marsh management.

Composition of hunters' bag.—Each marsh has its particular group of autumn waterfowl species, which makes up the hunter's bag. Usually a few kinds are dominant; others are shot rarely. Each year at Delta, since 1938, the staff of the Delta Station has examined hunters' bags to determine the species shot and the sex and age ratios of this kill. Between 1938 and 1941, Hochbaum (1944:133) examined 4,923 ducks shot by hunters. Between 1946 and 1950, an additional 6,716 ducks were examined by the Station staff. The results of Hochbaum's analysis of the species composition in the hunters' take were almost identical to those found by the later analysis. A combination of the data gathered by Hochbaum, and that gathered between 1946 and 1950, is shown graphically in Figure 44. Seventeen species made up the total bag during both periods, with 10 of these accounting for 99 per

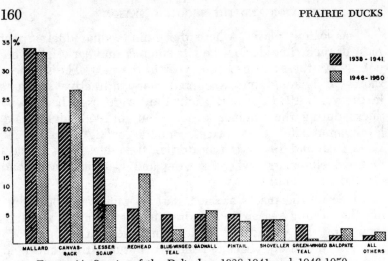

Figure 44. Species of the Delta bag 1938-1941 and 1946-1950.

cent of the total. Two species, the canvasback and the mallard, made up more than half (58 per cent) of the total kill.

During the period between 1946 and 1950, the ring-necked duck, bufflehead, American golden-eye, black duck, greater scaup, ruddy duck and white-winged scoter together made up only one per cent of the kill. Of these, only the ruddy duck and white-winged scoter were common nesting species at Delta.

An increase in the kill of ruddy ducks, following 1946, was noted. Although Hochbaum examined 4,923 ducks in hunters' bags between 1938 and 1941, he found only three ruddy ducks. In a sample of 6,716 ducks examined by the Station staff between 1946 and 1950, 24 ruddy ducks were found. This suggests an increased kill of this species in the later period. It seems logical to assume that the heavier gun pressure and lack of discrimination of hunters, which were apparent in the Delta marsh in the post-war years, may be responsible for these small differences in the bag. The ruddy had been ignored by experienced hunters which made up the bulk of the shooters involved in the earlier period.

As the hunting seasons progressed, changes in the composition of the hunters' bags were evident, and these probably represented changes in the marsh population. In the fall of 1946, for example, the hunting season opened on September 16 and continued for seven weeks, when the freeze-up forced the last ducks to leave. That fall we examined 5,208 birds in hunters' bags. The percentage of each species found is shown in Table 24. Both the blue-winged teal and the pintail were common at Delta in August and September, but departed in great numbers by the first of October. Of the autumn departure of the pintail from Delta, Hochbaum (1944:141) has said, "In June its numbers swell as drakes move to the marsh from other areas, and throughout the spring and early summer it is fully as common as the Mallard. After early August, however, the Pintail population, unlike that of the Mallard, dwindles." Of the blue-winged teal, Hochbaum (p. 140) says, "by late August there are bands of teal in every shallows; they attain peak numbers by early September, immediately prior to rapid and final disappearance. Blue-winged teal still are abundant during the first week of the shooting season, but by the last days of September or the early days of October, all but a few have gone."

Of all the more abundant species at Delta, the pintail and the blue-winged teal are the first to leave. Since later opening of the hunting seasons, which began in 1947 in Manitoba, a definite advantage has been given to these two species.

The mallard, the most abundant of all the autumn surface-feeding ducks, was represented heavily in hunters' bags each fall from opening day until the close of the season. The bags of gadwalls varied greatly with the source of the sample. The same was true of shovellers and green-winged teal. These species sometimes were common in hunters' bags taken in some parts of the marsh and entirely lacking in bags from the rest of the marsh, indicating a distinct preference on the part of the ducks for certain sloughs and potholes.

Comparisons of the duck kill by species during the periods 1938-1941 and 1946-1950 are shown graphically in Figure 44. The decrease in the number of blue-winged teal shot during the second period probably is caused by the later opening dates for hunting following World War II. I cannot account for the reduced kill of the green-winged teal during the later period. The lesser scaup, however, definitely showed a sharp reduction in populations moving through the Delta region in the years following World War II. This was evident in the reduced size of evening flights at freeze-up time.

TABLE 24. PERCENTAGE OF VARIOUS SPECIES OF DUCKS
FOUND IN HUNTERS' BAGS BY WEEKS OF SEASON, 1946

| Species | Percentage of total bag | | | | | | |
| | Week of season | | | | | | |
	1st	2nd	3rd	4th	5th	6th	7th
Mallard	40	23	30	26	42	49	54
Gadwall	7	9	3	4	7	14	0
Baldpate	3	5	2	2	5	0	0
Pintail	5	5	4	1	3	0	3
B-w teal	4	5	1	0	0	0	0
Shoveller	4	2	2	2	2	10	7
Redhead	17	21	15	6	6	5	1
Canvasback	16	27	38	49	20	12	11
Lesser Scaup	2	1	4	7	12	10	20
Others	2	2	1	3	3	0	4

"Old-timers" at Delta, who had hunted scaups for many years,
commented on the shortages of "bluebills." The decreased
scaup flights also were evident in the bag check. Hochbaum's
data shows that the scaup formerly shared the brunt of shooting pressure along with the canvasback and mallard, but
that this was less conspicuously so in more recent years. The
changes in the percentage of the total bag that scaups represented for 1938 and 1946 are shown in Figure 45, as are
the high points in the scaup bag for the years 1939, 1940 and
1941. The kill of this species always is low during the early
part of each shooting season, becoming highest about the
middle of October. During some years, the kill climbed to
a point where this species composed over half of the total
number of ducks bagged in any single week. In 1946, however, the scaup curve never climbed to over 20 per cent of
the bag.

Sex and age ratios.—Sex ratios of five of the most abundant
ducks at Delta are given in Table 25. These show the ratio
at the time of hatching, in the spring flight, and in the fall
hunters' bag. The sex ratio at hatching includes all birds that
could be sexed, which were raised in the Delta hatchery during these years. This is the secondary sex ratio, while the
ratios of grown birds tallied in the spring flight, and in hunt-

Figure 46. Well trained retrievers not only save crippled birds but also add to the sport of wildfowling.

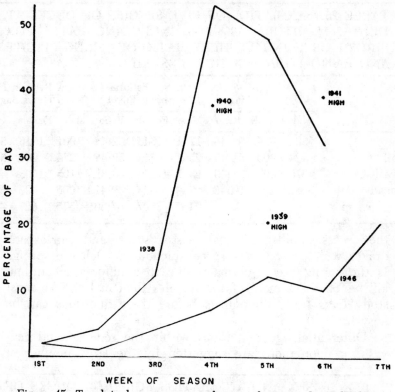

PERCENTAGE OF BAG

WEEK OF SEASON

Figure 45. Trend in lesser scaup numbers as shown in the Delta bag during five hunting seasons.

ers' bags, are tertiary sex ratios (May, 1939). These figures indicate that there was an almost even secondary sex ratio at hatching for the mallard, pintail and canvasback. Tertiary sex ratios for the pintail and mallard in the spring flight were almost even, while at the same time there was a heavy preponderance of males in the spring populations of canvasback, lesser scaup and redhead. Although these figures are the aggregate of 12 years' data, each individual year, taken by itself, also shows an excess of drakes.

Sex and age ratios for the autumn kill of the 10 most abundant ducks at Delta are shown in Table 26.

Some of the differences in the ratios between these fall data and the spring counts probably were caused by a differential movement of the two sexes. What happened to

TABLE 25. SEX RATIOS OF FIVE SPECIES OF DUCKS AT DELTA, MANITOBA, 1938-1950, INCLUDING RATIO FOR BIRDS RAISED IN HATCHERY, RATIO OF SPRING FLIGHT AND RATIO FOUND IN HUNTERS' KILL.

Species	Sex Ratio at Hatching			Sex Ratio in Spring Flight			Sex Ratio in Fall Hunters Bag		
	Male	Female	Ratio	Male	Female	Ratio	Male	Female	Ratio
Mallard	394	369	52:48	1751	1643	52:48	1901	1616	54:
Pintail	424	405	51:49	2567	2359	52:48	224	204	52:
Canvasback	315	307	51:49	2581	1229	68:32	1105	1508	42:
Redhead	342	294	54:46	749	524	59:41	606	495	55:
Lesser Scaup	...	...	....	7157	3507	67:33	575	627	48:

the excess numbers of male lesser scaup and canvasback drakes which always are so conspicuous at Delta in spring? Is the spring excess of males due to a differential autumn kill which takes more of the females than males? We must study bags from other areas before these questions can be answered.

Differential age migrations within the redhead and canvasback populations are suggested by the large number of

TABLE 26. SEX AND AGE RATIOS FOR 10,607 DUCKS OF TEN SPECIES EXAMINED IN HUNTERS' BAGS BETWEEN 1938 AND 1950.

Species	Male	Female	Ratio	Adult	Juv.	Ratio	Ratio adult female to juv.
Mallard	1901	1616	54:46	1279	2238	1:1.7	1:3.5
Gadwall	257	287	47:53	174	370	1:2.1	1:3.4
Baldpate	106	125	46:54	46	185	1:4.0	1:6.0
Pintail	224	204	52:48	173	255	1:1.5	1:2.7
G-w Teal	64	73	47:53	41	96	1:2.3	1:3.0
B-w Teal	199	220	47:53	96	323	1:3.4	1:4.5
Shoveller	224	191	54:46	97	318	1:3.3	1:5.3
Redhead	606	495	55:45	120	981	1:8.2	1:20.8
Canvasback	1105	1508	42:58	306	2307	1:7.5	1:11.0
Lesser scaup	575	627	48:52	305	897	1:2.9	1:6.5

juveniles shot, as compared with the number of adults; and this difference in the migration pattern between adults and juveniles is obvious in the blue-winged teal. Of 131 blue-winged teal checked in hunters' bags between 1946 and 1950, only 31 were adults. Banding records gave further indication that the adult blue-winged teal leave Delta ahead of the juveniles. Table 27 gives the age ratio of blue-winged teal trapped at the hatchery-pond during late summer and early fall. From this table, based on banding records of 1,039 birds trapped between August 1 and September 31 of these years, it is apparent that an early departure of adults precedes the mass migrations of the juveniles, or that the adults did not frequent the trap sites at Delta. Although most blue-winged teal left Delta before the hunting season, most of those that remained were juveniles.

That adult blue-winged teal drakes and adult drakes of other species depart before the juveniles has been shown by Hochbaum (1944:140), who wrote: "The adult blue-winged teal drake, like all other male river ducks except the mallard, departs from Delta soon after he has gained his new flight feathers in midsummer, and is uncommon in autumn."

Crippling loss from hunting.—The number of ducks crippled and lost by hunters long has been of concern to waterfowl biologists. That a large number of ducks which are hit survive has been shown by Elder (1950). Elder fluoroscoped 3,638 adult dabbling ducks in Manitoba and Saskatchewan and concluded (p. 502) that, ". . . one out of 4 drake mallards was carrying body shot. In this same sample, 1 out of 5 pintails, 1 out of 8 gadwalls, 1 out of 10 shovellers, and baldpates, and 1 out of 14 teal were carrying body shot."

TABLE 27. SEX AND AGE OF 1,039 BLUE-WINGED TEAL TRAPPED IN LATE SUMMER 1946, 1948, AND 1949

Period	Adult Drakes	Juvenile Drakes	Adult Hens	Juvenile Hens
August 1-15	16%	40%	7%	37%
August 15-31	1%	39%	1%	59%
September 1-15	0%	49%	0%	51%
September 15-31	0%	50%	0%	50%

In the course of examinations of hunters' bags, an attempt was made to gather information on the extent of loss of waterfowl through crippling. It soon was apparent, however, that questionnaires or direct questioning of hunters would not give a satisfactory figure. Many hunters were reluctant to give accurate figures. It was found that while some hunters were ashamed of their losses caused by crippling, other novice hunters were proud of the fact that they hit more birds than they bagged. This attitude was prevalent in the years immediately following World War II, when there was a noticeable increase of inexperienced and poorly equipped hunters in the marsh.

To ascertain the extent of crippling loss we obtained information from only those hunters whom we considered reliable. It was found that for 705 birds bagged, another 394 birds were lost as cripples. This ratio of about 0.5 of a cripple for each bird bagged applied only to the most conscientious hunters and probably was far below the crippling loss of the rank and file hunters in the Delta marsh.

The need for the use of trained retrievers to cut down the loss of cripples, and to add to the enjoyment of duck hunting, was pronounced in the Delta marsh, and in other areas. The number of hunters using retrievers in the prairie provinces of Canada is increasing rapidly. (Figure 46).

The pattern of autumn departure.—One was aware of the decreasing numbers of birds in the marsh as the season progressed. At the same time that flocks of birds were leaving, other flocks from the north were coming into the marsh; and one could not be certain which birds were newly arrived and which had been in the area for some time.

The ups and downs of the duck population in the Delta marsh, and the influx and departure of great flocks of birds, were predicted by experienced hunters in the area. The accuracy of their predictions could be judged at the boat-landings. Weeks of balmy weather meant poor hunting; and during "butterfly days" only novice hunters or jump-shooters went into the marsh. But chill north winds brought new quarry into the sky, and increased numbers of experienced hunters to the landings.

The freeze-up at Delta usually occurred suddenly and

with finality. Occasional temporary freeze-overs followed by warm days were recorded, however. It was noteworthy that the ducks appeared to advance and retreat with the temperature front. The disappearance of all birds from the marsh and the later reappearance of large numbers occurred in the fall of 1950, an event described to me by Hochbaum. That year an early freeze-up caused all ducks to leave early, but later warm weather brought ducks back into the marsh.

The dramatic final exodus of waterfowl from the breeding grounds may take place during one or two days. That the departure occurred over a wide front was evidenced in the fall of 1946. On November 2, a mass movement of lesser scaup was recorded in southern Manitoba. The exact time of arrival of the birds was noted in two places on latitude 50, 68 miles apart. Two independent observers, comparing their notes later, had obtained identical arrival times. Three other observers witnessed the flight at different points, but did not time its beginning.

The flight was seen at the village of Delta, at the south end of Lake Manitoba, by William Newman, Provincial Game Guardian, and by me. It began at 4:15 p.m., and it ended at sunset at 5:05 p.m. Albert Hochbaum, Torry Ward, and Roper Cadham all witnessed the flight at various other points in the marsh. The width of the flight recorded at Delta was 20 miles.

In the Netley marsh at the south end of Lake Winnipeg, 68 miles to the east, Mr. T. H. Schindler, Provincial Game Guardian, observed the migration and timed the flight. The coincidence of its arrival time with the Delta arrival time indicated a movement over a wide front.

At the same time that this movement of lesser scaup took place, about 500 whistling swans came from the northwest, down the lakeshore and into the big bays of the Delta marsh. Most of the mallards that had been in the vicinity had disappeared by the morning of the third.

The movement coincided with a wide-spread atmospheric condition. Inspection of the weather map showed a mass of cold air moving in from the west and north just behind the migration.

Some species of waterfowl are affected noticeably by cold

windy days in fall, but other species show little or no reaction. For example, a 4-day cold period on August 28, 29, 30, and 31, 1948, emptied the Delta marsh of blue-winged teal, but had no noticeable effect upon the pintail population. The mass departures of mallards and pintails that caused the downward trend in populations at Delta in 1948 (Figure 43), had no noticeable correlation with weather changes.

That not all spectacular mass departures in the fall coincide with rapidly falling temperatures and freezing has been shown by Rowan (1929), who recorded a mass departure of waterfowl from Alberta about ten days ahead of average. He sums up his observations by saying (p. 91), "The popular conception of the fall migration is a forced departure before winter storms. In contrast to this view, the southward movement herewith described, one of the most remarkable ever to have been observed in the Province of Alberta, took place during an unusually long spell of exceptionally high pressure with accompanying fine weather. The only wintry aspect was a gently falling thermometer . . . "

Thus we see that complex factors influence migration. At migration time the birds are physiologically ready to migrate. It seems likely that low atmospheric pressure and the advance of cold weather stimulate migration, but that migration will occur in the absence of low pressure and sudden cold.

SUMMARY

1. The first sign of a return to gregarious behavior among ducks in the summer was the presence of small groups of drakes that began to gather soon after the nesting season started.

2. These drakes did not necessarily remain in groups from the first gathering until fall; apparently some re-paired with hens that had lost their nests and isolated themselves with their hens for a renesting.

3. The passing of the nuptial plumage and the coming of the eclipse plumage corresponded, roughly, with the end of the breeding season for individual drakes.

4. The second group of birds to form gregarious flocks in the summer was made up of thwarted pairs. In samples counted at

Delta in 1950, it was found that this group comprised about 25 per cent of the early gathering groups.

5. In the vicinity of Delta, the molt of the primary feathers began about the last part of June. For individual ducks, the flightless period lasted about four weeks.

6. Many adult ducks came long distances to large marshes to spend their flightless period. But some adult hens were known to have nested near the fall gathering place.

7. Population build-ups at Delta began during the breeding season. The populations in Manitoba marshes usually reached their peak in mid-summer.

8. After the flightless period was over, large flocks of ducks rested on the sandbars, on grazed marsh edges, and made flights to grain fields to feed.

9. At Delta, grain field feeding was restricted to mallards and pintails. The habit of going to grain fields to feed apparently was learned over a period of years. Some groups of mallards were stubble feeders, some were not.

10. Predictable flights to grain fields occurred twice daily before dawn and at dusk, along definite flight lines. These lines of flight conformed to long winding creek beds that extended from the marsh into the prairie below. That there was some variation from these flights was evidenced by smaller numbers of feeding mallards and pintails on the grain fields at most any hour of the day.

11. Gunning, which occurred when the hunting season began, or the plowing of feeding areas, disrupted the regular stubble flights and diffused the flocks.

12. Stubble flights involved thousands of birds. Over 6,000 birds were counted in one flight that was checked until darkness made counting impossible.

13. Birds appeared to be well-fed after stubbling; their swollen crops were conspicuous. Upon returning to their loafing area, the birds alighted on the water, drank and slowly made their way up to bare ground where they sat or slept without exhibiting aggressive behavior.

14. When the usual loafing bars of these birds were destroyed by a rising lake level, the birds moved to the edges of the large bays.

15. Seventeen species of ducks constituted the hunters' kill in the Delta marsh. Ten of these made up 99 per cent of the

total kill. The canvasback and the mallard together composed 58 per cent of the kill.

16. As the hunting season progressed, species changes in the composition of the hunters' kill were evident, and reflected the changes in the duck population. Blue-winged teal and pintails were two species which departed early from the marsh and, though common in August and September, were nearly gone by October.

17. The flights of lesser scaup through the Delta marsh decreased noticeably in the years following World War II.

18. Figures on sex ratios indicated that there was an almost even secondary sex ratio at hatching for mallards, pintails and canvasback. Tertiary sex ratios for the pintails and mallards in the spring flight were almost even, while at the same time, there was a heavy preponderance of males in the spring population of canvasbacks, lesser scaups, and redheads.

19. It was found that, for 705 birds bagged by hunters, another 394 birds were lost as cripples. More widespread use of retrievers is recommended as a means of decreasing crippling loss, and increasing the value of the sport of wildfowling.

20. The picture of migration out of the marsh was confused by incoming migrants. However, differential age migration was apparent in some species. Blue-winged teal adults departed from the marsh ahead of the juveniles.

21. As freeze-up occurred, ducks appeared to advance and retreat with the temperature front.

22. As freeze-up approached, a mass migration was known to extend across a 20-mile front. The movement coincided with a widespread atmospheric condition.

23. Some species of waterfowl were more responsive to cold, windy weather than others. A 4-day cold period emptied the marsh of blue-winged teal, but produced no noticeable effect on pintails.

APPENDIX I

ACKNOWLEDGMENTS

*D*URING the field work, and the preparation of this manuscript, I received help from many people. Some contributed
new ideas, some gave useful suggestions, others assisted in finding
nests, banding birds, and doing many other tasks associated with
the study.

I wish to thank especially H. Albert Hochbaum, director of
the Delta Waterfowl Research Station for his part in initiating this
project, and his later help in all phases of field research and
manuscript preparation. To him also belongs the credit for the
black-and-white drawings which reflect his many years of experience in observing birds in the Delta marsh.

For technical advice, I thank also Dr. William H. Elder, Department of Zoology, University of Missouri, and Arthur S. Hawkins,
flyway biologist, U. S. Fish and Wildlife Service.

For their long hours in the field and faithful recording of data,
I am indebted to the three following summer assistants, each of
whom worked full time for one summer: Charles D. Evans,
University of Minnesota,[1] June-September, 1948; Eugene F. Bossen-
maier, University of Minnesota, May-September, 1949; and James
Houston, Colorado Agricultural and Mechanical College, June-
September, 1950. Others who gave me field assistance were:

---

[1] Organizational connections cited were effective at the time of participation.

George K. Brakhage, William R. Hecht, and Glen C. Sanderson of the University of Missouri; William H. Carrick, Royal Ontario Museum of Zoology; and J. Bernard Gollop, Canadian Wildlife Service.

I want to thank Alexander Dzubin, of the University of British Columbia, who furnished information on banded birds which returned to the Delta marsh in the summer of 1951.

The following pilot-biologists of the U. S. Fish and Wildlife Service took me on flights over the Delta marsh and adjoining areas to collect data for this study: Walter F. Crissey, Robert H. Smith, David L. Spencer, and Edward G. Wellein.

Residents of the village of Delta who assisted me by reporting nests and banded birds were: Raymond Burton, George Carmen, Frederick Caswell, John Douglas, William H. Hutchinson, Russell Loucks, Robert Loucks, Archie McDonald, Gordon McDonald, Percy Rutledge, Andrew Zurkon and others.

In the predation study made in 1939 and 1940, I was assisted by Edward Ward, Russell Ward, and Torrance Ward. For most of the hatchery work done to obtain young birds for release, and for use in the preparation of the embryo series, I am grateful to Peter Ward, superintendent of the Delta hatchery, and his assistant, Norman Godfrey.

For the use of camera equipment, I am grateful to James F. Bell of Minneapolis.

To the following people I am indebted for the use of unpublished data: Dr. Ira N. Gabrielson, president of the Wildlife Management Institute; Dr. Ernest E. Good, Department of Zoology, Ohio State University; Arthur S. Hawkins, flyway biologist, U. S. Fish and Wildlife Service; Merrill C. Hammond, biologist, U. S. Fish and Wildlife Service; Dr. Milton B. Trautman, Franz Theodore Stone Institute of Hydrobiology.

In the preparation of the manuscript, numerous helpful suggestions and criticisms were given me. I thank especially Dr. Joseph J. Hickey, Department of Wildlife Management, University of Wisconsin. I am indebted to Earl L. Atwood, Paul F. Hickie, E. R. Kalmach, Dr. Daniel L. Leedy, and Edward G. Wellein of the U. S. Fish and Wildlife Service; Dr. Robert A. McCabe, Department of Wildlife Management, University of Wisconsin; Mrs. Margaret M. Nice of Chicago; and Dr. William Rowan, Department of Zoology, University of Alberta.

For the drafting of charts and figures, I thank Grace M. Sowls. For financial assistance throughout the period of this project I am grateful to the North American Wildlife Foundation, and the Wildlife Management Institute, and to Dr. Ira N. Gabrielson and C. R. Gutermuth. I wish to thank the U. S. Fish and Wildlife Service for allowing me time to complete the manuscript. Permission to reproduce the frontispiece was granted by W. A. Murphy and Peter Curry of Winnipeg.

APPENDIX II

Common and Scientific Names of Plants Mentioned in Text
Nomenclature from Fernald (1950) and Bailey (1949)

Common cattail	*Typha latifolia*
Pondweed	*Potamogeton* sp.
Cordgrass	*Spartina pectinata*
Yellow cane, phragmites	*Phragmites communis*
Kentucky bluegrass	*Poa pratensis*
Whitetop	*Scolochloa festucacea*
Brome grass	*Bromus* sp.
Quackgrass	*Agropyron repens*
Skunk grass	*Hordeum jubatum*
Wheat	*Triticum aestivum*
Oats	*Avena sativa*
Barley	*Hordeum vulgare*
Hard-stem bulrush	*Scirpus acutus*
Soft-stem bulrush	*Scirpus validus*
Sedge	*Carex* sp.
Duckweed	*Lemna minor*
Willow	*Salix* sp.
Cottonwood	*Populus deltoides*
Bur oak	*Quercus macrocarpa*
Pigweed	*Chenopodium* sp.
Glasswort	*Salicornia* sp.
Wild mustard	*Brassica* sp.
Wild cherry	*Prunus serotina*
Flax	*Linum* sp.
Poison ivy	*Rhus toxicodendron*
Boxelder	*Acer negundo*
Red osier dogwood	*Cornus stolonifera*
Green ash	*Fraxinus pennsylvanica*
Bladderwort	*Utricularia* sp.
Snowberry	*Symphoricarpos occidentalis*
Red-berried elder	*Sambucus pubens*
Goldenrod	*Solidago* sp.
Aster	*Aster* sp.
Canada thistle	*Cirsium arvense*
Sow thistle	*Sonchus arvensis*

APPENDIX III

Common and Scientific Names of Animals Mentioned in the Text. Nomenclature from Anderson (1946), A.O.U. Checklist Ed. 4, (1931), Witherby, Jordan, Ticehurst and Tucker (1939), and Storer (1951).

NORTH AMERICAN BIRDS

Common Canada goose	*Branta canadensis canadensis*
Common mallard	*Anas platyrhynchos platyrhynchos*
New Mexican duck	*Anas diazi novimexicana*
Mottled duck	*Anas fulvigula maculosa*
Common black duck	*Anas rubripes tristis*
Gadwall	*Anas strepera*
Baldpate	*Mareca americana*
American pintail	*Anas acuta*
Green-winged teal	*Anas carolinensis*
Blue-winged teal	*Anas discors*
Shoveller	*Spatula clypeata*
Wood duck	*Aix sponsa*
Redhead	*Aythya americana*
Ring-necked duck	*Aythya collaris*
Canvasback	*Aythya valisineria*
Greater scaup duck	*Aythya neartica*
Lesser scaup duck	*Aythya affinis*
American golden-eye	*Bucephala clangula americana*
Bufflehead	*Bucephala albeola*
White-winged scoter	*Melanitta deglandi*
Ruddy duck	*Oxyura jamaicensis*
Marsh hawk	*Circus cyaneus hudsonius*
Hungarian partridge	*Perdix perdix perdix*
Bobwhite quail	*Colinus virginianus virginianus*
Sora rail	*Porzana carolina*
Coot	*Fulica americana americana*
Barn owl	*Tyto alba pratincola*
Snowy owl	*Nyctea scandiaca*
Domestic pigeon	*Columba livia*
Eastern crow	*Corvus brachyrhynchos brachyrhynchos*

Western crow	*Corvus brachyrhynchos hesperis*
Carolina chickadee	*Parus carolinensis carolinensis*
Eastern robin	*Turdus migratorius migratorius*
Eastern bluebird	*Sialia sialis sialis*
American redstart	*Setophaga ruticilla ruticilla*
Yellow-headed blackbird	*Xanthocephalus xanthocephalus*
Red-winged blackbird	*Agelaius phoeniceus phoeniceus*
Eastern song sparrow	*Melospiza melodia melodia*
Eastern snow bunting	*Plectrophenax nivalis nivalis*
Grebes	*Family Colymbidae*
Doves	*Family Columbidae*
Plovers	*Family Charadriidae*

EUROPEAN BIRDS

Grey lag-goose	*Anser anser anser*
Ringed plover	*Charadrius hiaticula*
Herring gull	*Larus argentatus*
English robin	*Erithocus rebecula*
European Jackdaw	*Corvus monedula*

NORTH AMERICAN MAMMALS

Short-tailed shrew	*Blarina brevicauda*
Snowshoe hare	*Lepus americanus*
Woodchuck	*Marmota monax*
Richardson ground squirrel	*Citellus richardsonii*
Franklin ground squirrel	*Citellus franklinii*
Pocket gopher	*Thomomys talpoides*
White-footed deermouse	*Peromyscus maniculatus*
Red-backed mouse	*Clethrionomys gapperi*
Field mouse	*Microtus pennsylvanicus*
Muskrat	*Ondatra zibethica*
Beaver	*Castor canadensis*
Jumping mouse	*Zapus hudsonius*
Long-tailed weasel	*Mustela frenata*
Mink	*Mustela vison*
Wolf	*Canis lupus*
Striped skunk	*Mephitis mephitis*
Elk	*Cervus canadensis*
White-tailed deer	*Odocoileus virginianus*
Bison	*Bison bison*

OTHERS

Grasshoppers	*Order Orthoptera*
Crickets	*Order Orthoptera*
Dragonflies	*Order Odonata*

OTHERS (Continued)

Beetles	*Order Coleoptera*
Flies	*Order Diptera*
Crayfish	*Class Crustacea*
Johnny Darter	*Family Percidae*
Toads	*Family Bufonidae*
Frogs	*Family Ranidae*
Skink	*Family Scinidae*

LITERATURE CITED

Anderson, Rudolph Martin. 1946. Catalogue of Canadian recent mammals. Bull. 102, Nat. Mus. Canada. 238 pp.

American Ornithologists Union. 1931. Check list of North American birds. Ed. 4, 526 pp. and supplements.

Bailey, L. H. 1949. Manual of cultivated plants. The MacMillan Co., New York. 1116 pp.

Barnes, William B. 1948. Unusual nesting behavior of a wood duck. Auk, 65:449.

Bellrose, Frank C. Jr. 1944. Duck population and kill, an evaluation of some waterfowl regulations in Illinois. Ill. Nat. Hist. Surv. Bull., 23 (Art. 2) :327-372.

......, and Elizabeth Brown Chase. 1950. Population losses in the mallard, black duck and blue-winged teal. Ill. Nat. Hist. Surv. Biol. Notes, 22:1-27.

Bent, Arthur Cleveland. 1923. Life histories of North American wildfowl, order Anseres (Part I). U. S. Nat. Mus. Bull. 126.

Bennett, Logan J. 1937. Grazing in relation to the nesting of the blue-winged teal. Trans. 2nd N. Amer. Wildl. Conf. :393-396.

....1938. The blue-winged teal, its ecology and management. Collegiate Press, Inc., Ames, Iowa. 144 pp.

Bird, Ralph D. 1930. Biotic communities of the aspen parkland of central Canada. Ecology, 11:356-442.

Brakhage, George K. 1953. Migration and mortality of ducks hand-reared and wild-trapped at Delta, Manitoba. Jour. Wildl. Mgt., 17:465-477.

Bue, I. G., Lytle Blankenship and William H. Marshall. 1952. The relationship of grazing practices to waterfowl breeding populations and production on stock ponds in western South Dakota. Trans. 17th N. Amer. Wildl. Conf. :396-414.

Buss, Irven O., Ronald K. Meyer and Cyril Kabat. 1951. Wisconsin pheasant reproduction studies based on ovulated follicle technique. Jour. Wildl. Mgt., 15:32-46.

Cartwright, B. W. 1944. The "crash" decline in sharp-tailed grouse and Hungarian partridge in western Canada and the role of the predator. Trans. 9th N. Amer. Wildl. Conf. :324-330.

....1945. Some results of waterfowl banding in western Canada by Ducks Unlimited (Canada). Trans. 10th N. Amer. Wildl. Conf. :332-338.

Delacour, Jean and Ernst Mayr. 1945. The family Anatidae. Wilson Bull., 57:3-55.

Dice, Raymond Lee. 1920. Notes on some birds of interior Alaska. Condor, 22:176-185.

Dixon, Joseph S. 1943. Birds observed between Point Barrow and Herschell Island on the arctic coast of Alaska. Condor, 45:49-57.

East, Ben. 1930. My friends of the sycamore. Bird-Lore, 32(1):4-7.

Elder, William H. 1950. Measurement of hunting pressure in waterfowl by means of X-ray. Trans. 15th N. Amer. Wildl. Conf. :490-504.

Emlen, John T. 1942. Notes on a nesting colony of western crows. Bird-Banding, 13:143-153.

Engeling, Gus A. 1949. The mottled duck—a determined nester. Texas Game and Fish, 7(8):6-7.

Errington, Paul L. 1942. On the analysis of productivity in populations of higher vertebrates. Jour. Wildl. Mgt., 6:165-181.

.......1946. Predation and vertebrate populations. Quart. Rev. Biol., 21:144-177.

Evans, Charles D. 1951. A study of the movements of waterfowl broods in Manitoba. Unpub. thesis. University of Minnesota. 134 pp.

Fabricius, E. 1951. Zur Ethologie junger Anatiden. Acta Zoologica Fennica, 68:1-175.

Farley, Frank L. 1932. Birds of the Battle River region of central Alberta. Inst. Applied Art. Ltd., Edmonton. 85 pp.

........1939. Further notes on the bird-life of Churchill, Manitoba, Can. Field Nat., 48:57.

Farner, Donald S. 1945. The return of robins to their birthplaces. Bird-Banding, 16:81-99.

Fernald, Merritt Lyndon. 1950. Gray's manual of botany. Eighth Ed. American Book Co. New York, Cincininati, Chicago. 1632 pp.

Furniss, O. C. 1938. The 1937 waterfowl season in the Prince Albert District, Central Saskatchewan. Wilson Bull., 50:17-27.

Girard, George L. 1939. Notes on the life history of the shoveller. Trans. 4th N. Amer. Wildl. Conf. :364-377.

.......1941. The mallard: its management in western Montana. Jour. Wildl. Mgt., 5:233-259.

Hawkins, Arthur S. and F. Graham Cooch. 1948. Waterfowl breeding conditions in Manitoba, 1948. In Waterfowl populations and breeding conditions, summer 1948. Spec. Scient. Rep. No. 60, U. S. Dept. Int., Fish & Wildlife Service, Washington, D. C., :76-98.

Heit, William S. 1948. Texas coastal waterfowl concentration areas and their 1947-48 wintering populations. Trans. 13th N. Amer. Wildl. Conf. :323-338.

Hickey, Joseph J. 1943. A guide to bird watching. Oxford, London, New York and Toronto. 262 pp.

.......1952. Survival studies of banded birds. Spec. Scient. Rep.: Wildlife No. 15, U. S. Dept. Int., Fish & Wildlife Service, Washington, D. C. 177 pp.

.......1953. A review of: The relationship of grazing practices to waterfowl breeding populations and production on stock ponds in western South Dakota by I. G. Bue, Lytle Blankenship and William H. Marshall. Trans. 17th N. Amer. Wildl. Conf.: 396-1 *In* Auk, 70:224-225.

Hochbaum, H. Albert. 1944. The canvasback on a prairie marsh. Amer. Wildlife Inst., Washington, D. C. 201 pp.

.......1946. Recovery potentials in North American waterfowl. Trans. 11th N. Amer. Wildl. Conf. :403-418

.......1947. The effect of concentrated hunting pressure on waterfowl breeding stock. Trans. 12th N. Amer. Wildl. Conf. :53-62.

Holm, E. O. 1947. Sexual behaviour and seasonal changes in the gonads and adrenals of the mallard. Proc. Zool. Soc. of London, 117:281-304.

Howard, H. Eliot. 1920. Territory in bird life. Murry, London. 308 pp.

Jull, Morley A., Morley G. McCartney and Hussein M. El-Ibiary. 1948. Hatchability of chicken and turkey eggs held in freezing temperatures. Poultry Sci., 27:136-140.

Kalmbach, E. R. 1937. Crow-waterfowl relations. U. S. Dept. Agr. Circular 443.

.......1938. A comparative study of nesting waterfowl on the Lower Souris Refuge: 1936-1937. Trans. 3rd N. Amer. Wildl. Conf.: 610-632.

.......1939. Nesting success: its significance in waterfowl reproduction. Trans. 4th N. Amer. Wildl. Conf.: 591-604.

Kendeigh, S. Charles. 1947. Bird population studies in the coniferous forest biome during a spruce budworm outbreak. Dept. Lands and Forests, Ontario, Canada, Biol. Bull., 1. 100 pp.

Kortwright, Francis H. 1943. The ducks, geese and swans of North America. Amer. Wildlife Inst., Washington, D. C. 476 pp.

Lack, David. 1940. The age of the blackbird. Brit. Birds. 36:166-175.

.......1943. The life of the robin. H. F. and G. Witherby Ltd., London. 224 pp.

. . . . . . .1947. The significance of clutch-size. Ibis, Apr. 1947:302-353.

Laskey, Amelia R. 1940. The 1939 nesting season of bluebirds at Nashville, Tennessee. Wilson Bull., 52:183-190.

Laven, H. 1940. Ueber Nachlegen und Weiterlegen. Ornith. Monatsber, 48:131-136.

Leedy, Daniel L. 1950. Ducks continue to nest after brush fire at Castalia, Ohio. Auk, 67:234.

Leopold, Aldo. 1933. Game management. Charles Scribners Sons, New York, London. 481 pp.

Lincoln, Frederick C. 1939. The migration of American birds. Doubleday, Doran, New York. 189 pp.

Lindsey, Alton A. 1946. The nesting of the New Mexican duck. Auk, 63:483-492.

Lorenz, Konrad Z. 1937. The companion in the birds' world. Auk, 54:245-273.

. . . . . . .1941. Vergleichende Bewegungsstudien an Anatinen. Jour. fur Ornith., 89, Erg. Bd. 3, Sonderh., :194-294.

. . . ., and N. Tinbergen. 1937. Taxis und Instinkthandlung in der Eirollbewegung der Graugans. I. Zeitschr. f. Tierpsychol., 2:1-29.

Low, Jessop B. 1945. Ecology and management of the redhead (Nyroca americana) in Iowa. Ecol. Monog., 15(1):35-69.

Mayr, Ernst. 1935. Bernard Altum and the territory theory. Proc. Linnaean Soc., N. Y. 45-46:24-38.

. . . . . . .1939. The sex ratio in wild birds. Amer. Naturalist, 73:156-179.

. . . . . . .1942. Systematics and the origin of species. Columbia Univ. Press, New York. 334 pp.

McCabe, Robert A. 1947. The homing of transplanted young wood ducks. Wilson Bull., 59:104-109.

Millais, J. G. 1902. The natural history of the British surface-feeding ducks. Longmans, Green and Co., London.

Munro, J. A. 1943. Studies of waterfowl in British Columbia: mallard. Can. Jour. Res., 21:223-260.

. . . . . . .1944. Studies of waterfowl in British Columbia: pintail. Can. Jour. Res., 22(D):60-86.

Nice, Margaret Morse. 1937. Studies in the life history of the song sparrow. I. Trans. Linnaean Soc., N. Y., 4. 247 pp.

. . . . . . .1941. The role of territory in bird life. Am. Mid. Nat., 26:441-487.

. . . . . . .1943. Studies in the life history of the song sparrow. II. Trans. Linnaean Soc., N. Y., 6. 328 pp.

. . . . . . .1953. Some experiences in imprinting young ducklings. Condor, 55:33-37.

Noble, G. K. and D. S. Lehrman. 1940. Egg recognition by the laughing gull. Auk, 57:22-43.

Oates, W. Coape. 1905. Wild ducks, how to rear and shoot them. Longmans, Green and Co., London.

Odum, E. P. 1942. Long incubation by a Carolina chickadee. Auk, 59:430-431.

Phillips, John C. 1928. The crow in Alberta and the work of the American wildfowlers. Trans. 15th Nat. Game Conf.:146-155.

Porsild, A. E. 1943. Birds of the McKenzie delta. Can. Field Nat., 57:19-35.

Riddle, Oscar. 1916. Studies on the physiology of reproduction in birds. I. The occurrence and measurement of a sudden change in the rate of growth of avian ova. Amer. Jour. Physiology, 41(3):337-396.

Rowan, William. 1929. Migration in relation to barometric and temperature changes. Bull. of the Northeastern Bird-Banding Assoc., 5:85-92.

Seton, Ernest Thompson. 1929. Lives of game animals. Doubleday Doran Co., New York. 4 vol.

Smith, Robert H. and Arthur S. Hawkins. 1948. Appraising waterfowl breeding populations. Trans. 13th N. Amer. Wildl. Conf.: 57-62.

Sowls, Lyle K. 1948. The Franklin ground squirrel (*Citellus franklinii*) and its relationship to nesting ducks. Jour. Mamm., 29:-113-137.

. . . . . . .1949. A preliminary report on renesting in waterfowl. Trans. 14th N. Amer. Conf. :260-275.

Stieve, H. 1918. Die Entwicklung des Eierstockeies der Dohle (Coloeus mondedula). Arch. mikr. Anat. 92. Ab. II:137-288.

Storer, Tracy I. 1951. General zoology. McGraw-Hill Book Co., Inc., New York, Toronto and London. 832 pp.

Stoudt, Jerome H. and Floyd H. Davis. 1948. 1948 Waterfowl breeding ground survey north central region. *In* Waterfowl populations and breeding conditions, summer 1948. Spec. Scient. Rep. No. 60, U. S. Dept. Int., Fish & Wildlife Service, Washington, D. C. :123-148.

Stresemann, Erwin. 1940. Zeitpunkt und Verlauf der Mauser bei einigen Entenarten. Jour. fur Ornith. 88:288-333.

Swanson, Gustav. 1926. A persistent bluebird. Bird-Lore, 28:339.

Thomas, Ruth Harris. 1946. A study of eastern bluebirds in Arkansas. Wilson Bull. 58:143-183.

Tinbergen, N. 1939. The behavior of the snow bunting in spring. Trans. Linnaean Soc., N. Y., 5. 95 pp.

. . . . . . .1951. The study of instinct. Oxford, London. 228 pp.

Trautman, Milton B. 1949. Observations on the spring courtship behavior of the black duck. Unpub. ms. Read before the 30th Annual meeting of the Wilson Ornith. Club on April 22, 1949. For abstract see Wilson Bull., 61:201.

Valikangas, I. 1933. Finnische Zugvogel aus englischen Vogeleiern. Vogelzug, 4:159-166 (Cf. Margaret M. Nice 1934, Bird-Banding, 5:95).

Van Den Akker, John B. and Vanez T. Wilson. 1950. Twenty years of bird-banding at Bear River migratory bird refuge, Utah. Jour. Wildl. Mgt., 13:359-376.

Ward, Edward. 1942. Phargmites management. Trans. 7th N. Amer. Wildl. Conf. :294-298.

Watson, John B. and K. S. Lashley. 1915. An historical and experimental study of homing. Papers from the Dept. of Marine Biology., Carnegie Inst. of Washington, 7:9-60.

Williams, Cecil S. and William H. Marshall. 1938. Duck nesting studies, Bear River migratory bird refuge, Utah, 1937. Jour. Wildl. Mgt., 2:29-48.

Witherby, H. F., F. C. R. Jourdain, Norman F. Ticehurst, and Bernard W. Tucker. 1939. The handbook of British birds. 3. H. F. and G. Witherby, London. 387 pp.

Wormald, H. 1910. The courtship of the mallard and other ducks. Brit. Birds, 4:2-7.

INDEX

Acknowledgments, 171.
Agassiz, Lake, 2.
age ratio, 162-165.
aggressiveness (*see* behavior).
agricultural practices and management, 77.
Alaska, 22.
Alberta, 114, 122, 123, 125, 155, 168.
ANACREON, 25.
arrival (*see* spring arrival).
artificial stocking, 43.
Assiniboine River, 19.
autumn,
   behavior and shooting season, 151-170.
   pattern of departure, 161, 166.
badger, 71.
bag,
   composition of hunters' (Delta), 159-165.
   marsh area preference by some species, 161.
baldpate,
   courtship, 22.
   eggshell carrying, 107.
   increasing in numbers in July at Delta, 154.
   live birds carrying body shot, 165.
   mobility of broods, 144.
   nesting, 4.
   place in hunter's bag, 160, 162.
   prenuptial courtship, 98, 99.
   sex and age ratio in bag, 164.
   sitting on dead eggs, 96-98.
   species association, 23.
   spring courtship, 22.
   spring migration, 11.
bands, 1, 7.
barley, 3, 66.
   wild, 69, 70.
BARNES, 130.
barometric pressure and migration, 17.
behavior,
   absence of aggressive behavior, 59, 60.
   aggressiveness and nesting chronology, 59.
   aggressiveness and territory, 52.
   attachment to home range, 49, 50.

autumn behavior, 151-170.
brood behavior at hatching time, 144-148.
courtship, 98-101.
defense, 52, 57, 59, 60, 61.
defense flights, 16, 57, 59.
desertion of nest, 98.
display, 21, 99.
dissolution of pair status, 96.
drake, accompanying the hen, 40, 41, 95, 96.
drake, circling and calling over nest site, 95, 96.
drake's reaction to stuffed dummies, 60, 61.
drakes, teasing by hens, 99.
effect of temperature on, 85-89.
egg recognition, 103.
eggshell carrying, 103-108.
evening flights of pairs, 91-93.
feigning by hen, 148.
flights, territorial vs. courtship, 99.
gathering of "thwarted" pairs, 138, 152-153.
hen and brood behavior, 143-150.
hen's approach to the nest, 5, 106.
hen's defense of brood, 149.
hen hiding brood, 147, 148.
hen hiding with brood, 148.
laying hen, 6.
nesting (*see* nesting behavior).
of fall flocks, 156-159.
of paired mallards and pintails, 21.
pursuit of teasing hens, 151.
renesting courtship, 98-101.
retrieving of displaced eggs, 101, 102.
selecting nest site, 13.
sexual (of transients), 22.
sharing of loafing area, 56, 57.
sitting on dead eggs, 96-98.
teasing hens, 99.
tolling by hens, 147-148.
of transients, 21.
BELLROSE and CHASE, 29, 31, 32, 33.
BENNETT, 71, 102, 119, 129.
BENT, 22.
BIRD, 96.
black duck,
   place in Delta bag, 160.
   sitting on dead eggs, 97.

survival rate, 29.
territorial behavior of, 60.
bladderwort, 72.
bluebird, eastern, 97.
bluegrass, 65-69.
blue-winged teal, (*see* specific reference).
botulism, 113.
box-elder, 68.
BRAKHAGE, 43.
breeding season (*see* nesting season).
breeding tradition, 41.
British Columbia, 28, 29, 59.
brome grass, 66, 68.
broods,
behavior on hatching, 146, 147.
feigning by hen, 148.
habitat, 149.
hen and brood behavior, 143.
hen's defense of, 149.
hiding by hen, 147, 148.
hiding with hen, 148.
movements, 144.
reactions to calls of the hen, 144-147.
renesting after loss of, 136-137.
BUE, BLANKENSHIP, and MARSHALL, 70.
bufflehead,
place in Delta bag, 160.
bulrush, 1, 2, 66, 69, 70, 72, 149.
bur oak, 68.
"burned out" marshes, 41.
burning (*see* fire).
to control undesirable vegetation, 76.
marsh firelanes, 78.
selective and early to aid waterfowl, 77, 78.
Canada thistle, 3, 66, 69.
canvasback,
ability to pioneer, 42.
differential age migration, 164-165.
habitat, 1.
mobility of broods, 144.
nest building, 93-95.
place in hunter's bag, 160, 162.
rest and feeding periods, 102, 103.
sex and age ratio in bag, 164.
sex ratio at hatching, 164.
sex ratio in spring population, 164.
CARTWRIGHT, 40, 130.
cattail,
in the Delta marsh, 1, 2.
destruction of improves edge, 70.
plants of permanent water areas, 72.
reaction to changing water conditions, 69, 70.

cattle (*see* grazing),
trampling of, and loafing edge for ducks, 70.
trampling of, and nest mortality, 113.
census methods, 56, 138.
chickadee, Carolina,
sitting on dead eggs, 97.
chronology,
events in life of a shoveller, 51.
nesting, 59, 83, 84, 86, 87.
clutch size,
criterion for distinguishing renests, 130.
determination of, 136.
first nests vs. renests, 130-132.
reduction of, 132.
color-marking, 6, 7.
Columbia, 1.
continuous laying, 134-136.
coots,
buffer between mink and ducks, 119.
killed by striking telephone wires, 117.
cordgrass, 3, 65, 66, 69, 70.
cottonwood, 68, 70.
courtship,
aerial prenuptial of transients, 22.
baldpate, 22.
effect of temperature on, 87.
green-winged teal, 22.
nuptial, 98, 99.
of transients, 21.
pintail courtship display, 21.
prenuptial, 98.
renesting, 98-101.
crippling loss, 165-166.
crow,
change in density, 122, 123.
crow control, 125, 126.
crow-waterfowl relationship, 113-115.
foods of, 118-120, 122.
human disturbance and crow predation, 115.
movements of and nest predation, 124, 125.
sitting on dead eggs, 97.
renesting, 98-101.
Cuba, 1.
decimating factors, 113.
deer, white-tailed, 71.
defended area,
and territory, 50-54, 59.
defense flights, 16, 57, 59.
peak of, and beginning of nesting, 57.

of gadwall from shared jumping-off place, 57.
(*see* also behavior).
vs. pursuit flights, 52.
DELACOUR and MAYR, 52, 99.
determinate vs. indeterminate layers, 129.
DICE, 22.
differential age migrations,
in redhead, canvasback, and teal, 164-165.
differential movements,
of sexes, 163.
display,
prenuptial, 99.
absence of mallard display in transients, 21.
pintail courtship display of transients, 21.
distances between nests, 33, 34.
from nests to water, 73, 74.
disturbance,
toleration by hens, 98, 108.
diving ducks (*see* species).
abortive migrations, 19.
habitat, 1.
and pioneering, 42.
sex ratio, 23, 164.
DIXON, 22.
dogs,
reaction of ducks to, 147.
to reduce crippling loss, 166.
use of, 4, 5.
dogwood, 68.
domestic fowl,
growth of ova, 136.
down, 89, 132.
drakes,
attachment to home range, 49, 50, (*see* also behavior).
homing of, 40, 41.
second drakes for renesting, 50.
drought, 113.
ducklings (*see* broods).
Ducks Unlimited (Canada).
crow control, 125.
fire control and education, 78.
duckweed, 72.
EAST, 97.
eggs,
aging of embryos, 82, 83.
dropping of, 87.
loss of, 109.
loss from freezing, 89.
rate of laying, 95.
recognition, 103.
retrieving of displaced eggs, 101, 102.
sitting on dead eggs, 96-98.
temperature and hatchability, 87.
time of day for laying, 95.
wooden eggs, 103.
eggshells,
carrying of, 103-109.
in dried up sloughs, 108.
in hatched nests, 108.
induced eggshell carrying, 106, 107.
ELDER, 165.
elder, red-berried, 68.
embryos,
aging of, 82, 83.
killed by fire, incubation of, 96-98.
EMLEN, 97.
ENGELING, 130.
England,
transplanted eggs and juvenile return, 35.
English robin,
and territory, 52-53.
ERRINGTON, 123, 130.
Europe and Asia,
fall gatherings of ducks far from breeding marshes, 154.
EVANS, 144, 149.
evening flights,
flights of scaup at freeze-up time, 167.
interval of time between evening flights and first nesting, 92, 93.
of pairs, 91-93.
experimental homing,
vs. migrational homing, 25.
FABRICIUS, 146-147.
fallow fields,
as nesting terrain, 66-67.
FARLEY, 103, 122, 155.
FARNER, 35.
feeding, 20, 21, 54, 102, 103, 155-159.
feigning (*see* behavior).
Finland,
homing of juvenile mallards from transplanted eggs, 35.
fire,
cause of nest loss, 113.
and incubation of dead eggs, 96.
and management, 78.
flax, 2, 66, 77.
flightless period,
duration, 153-154.
flock formation for, 151.
flights, defense, 16, 57, 59.
flights, evening, 91, 92.
floating logs, use of, 1, 8, 149.
flock size,
during migration, 23.
during stubble feeding, 158.

flooding and nest loss, 113, 114.
flushing bar, use of, 4.
fog,
    effect on migration, 20.
follicles,
    laying indicator in ducks, 100.
    regression of, when incubation
        begins, 135-136.
    resorption and rebuilding of, 133,
        136.
    size in hen exhibiting evening
        flight behavior, 92.
"following" reaction (see
        FABRICIUS).
freeze-up, 160, 166-167.
    evening flights of scaup at
        freeze-up, 160, 166-167.
    influence on ducks, 166-167.
freezing of wild eggs, 88, 89, 97.
FURNISS, 114.
GABRIELSON, 22.
gadwall (see specific reference).
Germany, 138.
GIRARD, 102, 114.
glasswort, 3.
goldeneye, American,
    place in Delta bag, 160.
goldenrod-aster, 66.
gonads,
    and plumage change, 152.
GOOD, 124-125.
goose, Canada,
    artificial stocking of, 43.
    homing of and evolution, 25-26.
    loss of eggs by freezing, 88, 89.
    migration of, 11, 17.
goose, grey,
    imprinting and the parent-
        companion, 146.
    retrieving of eggs, 101.
gopher, pocket,
    abundance, 71.
    influence on nesting cover, 71.
grain stubble (see stubble).
grazing,
    to benefit waterfowl, 78, 159.
    grazed pastures on the study
        area, 67.
    grazing densities desirable for
        ducks, 70-71.
    improves loafing edge for ducks,
        70.
    influence on nesting cover, 69-71.
    relationship to badger, skunk, and
        squirrel, 71.
gunning pressure, 41, 160-161.
habitat, 1, 36, 213.
HAMMOND, 93-94, 97.
hatchability and temperature, 88.

hatching,
    incubator vs. wild hatching,
        143-144.
hawk, marsh,
    foods of, 119-121, 128.
    predator on ducklings, 113.
HAWKINS, 21, 149.
HAWKINS and COOCH, 154.
hay (wild),
    cutting of and nesting cover, 67.
    marsh edge crop, 77.
HEINROTH, 99, 146.
HEIT, 22.
HICKEY, 12, 25, 29, 39, 43.
HOCHBAUM, 19, 21, 23, 32, 41, 42,
        52, 53, 81, 93, 98, 99, 104, 123,
        130, 148, 152, 153, 154, 155,
        159, 161, 171.
HOLM, 152.
home range, 47-63.
    duration of home range
        attachment, 49-50.
    an example of, 48, 49, 51.
    in mammals, 47.
    and territory, 47-63.
    in waterfowl, definition, 48.
homing (see migrational homing).
homing instinct, 34.
homing pigeons, 25.
HOWARD, 50.
hunters' bag, composition of, 162,
        163, 164.
hunting,
    and crippling loss, 165-166.
    influence on stubble feeding, 156.
    of breeding waterfowl, 41-42.
hunting pressure (see gunning
        pressure).
hunting seasons,
    effects of later, 41-42, 160-161.
identification (see bands, paint,
        marking).
Illinois, 29, 35.
"imprinting", 145-147.
incitement call, 99-100.
incubation,
    of infertile or dead eggs, 96-98.
    regression of follicles and
        incubation, 135-136.
    renesting interval, 132-134.
interspersion,
    definition of, 73.
    at Delta, 73.
isotherms, 17.
jackdaw, 133.
JULL, McCARTNEY and
        EL-IBIARY, 88.
juveniles,
    in hunters' kill, 164.

nesting place of returning juvenile
  hens, 38, 39.
and pioneering, 41-42.
release and return of captive-
  reared, 34-41.
return of, 40-41.
KABAT, BUSS and MEYER, 100.
KALMBACH, 114, 115, 123, 125,
  126, 127, 130.
KENDEIGH, 53.
kill (see bag),
  relationship to return at Delta,
  32-33.
KORTWRIGHT, 26.
LACK, 29, 52, 130.
land use and nesting cover, 66-69,
  77-78.
land-water pattern, 73, 75.
LASKEY, 97.
LAVEN, 133, 138.
laying rate, 95.
lead shot in live birds, 165.
LEEDY, 97.
LEOPOLD, 73, 97, 98.
LINCOLN, 17, 34, 41.
LINDSEY, 104.
literature cited, 179.
"loafing",
  activity of transients, 21.
  loafing of broods, 149.
  loafing of fall flocks, 159.
  species association, 23.
"loafing" places, 7, 8, 20, 21.
  and behavior of shoveller pair, 54.
  necessary to waterfowl, 159.
  trading of loafing places by pairs,
  56, 57.
LORENZ, 99, 145.
LORENZ and TINBERGEN,
  101-102.
Louisiana, 22.
LOW, 93, 102, 119, 129, 131, 138.
LYNCH, 22.
mallard (see specific reference).
mammals (see species),
  home range in, 47.
  influence on nesting cover, 71.
management,
  and agricultural practices, 77-78.
  and grazing (see grazing also), 69.
  and homing, 41-44.
  and marsh fire, 78.
  and pioneering, 43.
  and predator control, 125-127.
  renesting and inventory counts,
  138-139.
Manitoba, Lake, 2, 16, 68, 159.
marking birds for identification,
  1, 6-8.
MAYR, 25, 50, 163.

McCABE, 35.
MacKenzie River delta, 22.
Michigan, 43.
migration,
  abortive attempts of, 19-20.
  autumn, 166-168.
  distance traveled and time
  involved, 22-23.
  flock size, 23.
  and fog, 20.
  main flight, 13.
  reversed migration, 18-19.
  species association, 23.
  spring, 11-24.
  and the weather, 17-19.
migrational homing, 25-45.
  adult vs. juvenile hens, 37.
  of captive reared juveniles, 34-41.
  definition of, 25.
  of drakes, 40.
  homing rate vs. calculated survival
  rate, 26, 31, 33.
  of juvenile drakes, 36, 40.
  of mallards vs. pintail and teal, 36.
  of pair to hen's natal marsh, 40-41.
  to hatching place in Finland, 35.
  vs. experimental homing, 25.
  and waterfowl management, 41-44.
  of wood ducks, 35.
milfoil, water, 72, 151.
MILLAIS, 99.
mink,
  foods of, 119, 122.
  hen's defense of brood against, 149.
  inhabitant of nesting meadows, 71.
  as predator on ducklings, 113.
molt,
  duration of 153-154.
molting areas,
  location of, 154.
mortality,
  adult, 116-117.
  of blue-winged teal vs. mallard, 32.
  foods of predators, 117-122.
  from striking telephone wires,
  116-117.
  nest, 113-116.
nest loss because of flooding, 115-116.
  rates, 28-29, 32.
mottled duck,
  renesting of, 130.
mouse, meadow, 71.
movement,
  of broods, 144.
  of duck population during the day,
  54-57.
  during late summer, 73.
mowing,
  to control undesirable vegetation,
  76.

MUNRO, 28, 40, 59, 155.
nest,
  built-up nests, 116.
  canopy, 94.
  construction of, 93-94.
  desertion, 98.
  determination of starting date of,
    82.
  distances between nests of
    individuals, 33, 34, 137.
  distances from nests to water, 73,
    74.
nest bowl,
  construction of, 93.
nest loss, 113-114.
  compensated for by renesting,
    127, 144.
  variation in early-season nest
    losses, 144.
nest moving,
  response to, 108-109.
nest, scent of, 4.
nest trap, 5, 6.
nesting behavior,
  building the nest, 93-94.
  disappearance of eggs, 109.
  dissolution of pair status, 96.
  drake, accompanying the hen,
    95-96.
  egg recognition, 103.
  eggshell carrying, 103-108.
  incubating hen sitting on dead eggs,
    96-98.
  nest desertion, 98.
  renesting courtship, 99-100.
  response to nest moving, 108-109.
  rest and feeding periods, 102-103.
  retrieving displaced eggs, 101-102.
  selection of the nest site, 91-93.
nesting chronology,
  and defense flights, 57, 59.
  of mallard, pintail, and blue-
    winged teal, 83-84, 87.
  of shoveller and gadwall, 87.
nesting cover,
  and grazing, 69.
  influence of wild mammals, 71.
  phragmites, 68.
  preferred plants, 65-66.
  use of stubble fields, 66-67.
nesting season, 81-90.
  comparison of 1949, 1950 nesting
    seasons, 86.
  determining the end of, 89.
  effect of weather on beginning and
    length of, 84-89.
  length of, 84.
  mortality, 113-116.
nesting success, 114.

nesting terrain, 65-79 (*see* nesting
    cover).
  agriculture and waterfowl
    management, 77-78.
  components of, 65.
  control of undesirable vegetation,
    76-77.
  distances from nests to water,
    73-74.
  dry and wet years, influence of, 77.
  influence of grazing on nesting
    cover, 69-71.
  influence of water levels on nesting
    cover, 68-69.
  interspersion, 73.
  land use and nesting cover at
    Delta, 66-68, 77-78.
  land-water pattern, 73, 75.
  quality of nesting areas, 75-76.
NICE, 17, 34, 52, 133, 146.
NOBLE and LEHRMAN, 101.
North Dakota, 94, 97, 114.
OATES, 104.
oats, 3, 66, 155.
ODUM, 97.
ovulation,
  in pheasants, 100.
  in waterfowl, 100.
owl, barn, 97.
owl, great-horned, 113.
paint, use of, 6-7.
pair formation,
  behavior of pairs, 21.
  dissolution of pair status, 96.
  time of, 21.
Panama, 1.
partridge, Hungarian,
  sitting on dead eggs, 97.
PHILLIPS, 122, 125.
phragmites,
  in the bays and sloughs at Delta, 2.
  control of, 76.
  destruction of improves edge, 70.
  preference rating for duck nests,
    66, 68.
  reaction to changing water
    conditions, 70.
pigeons (*see* homing pigeons).
pigweed, 3.
pintail (*see* specific reference).
pioneering, 42-43.
plant preferences,
  for duck nests, 66.
plant succession, 69.
plover, ringed, 133, 138.
plowing,
  influence on stubble flights, 157.
  and nest losses, 67.

plumage,
  aid to interpreting actions of
    ducks, 152.
  duration of molt, 153, 154.
  plumage change and breeding
    status, 152.
plumage marking, 1, 6-8.
Poland, 133.
pondweeds, 2, 72, 151.
population,
  build-up in late summer, 89, 154,
    156.
  changes in population reflected in
    bang, 160-161.
  populations and weather, 167.
  variation in ditch population,
    54-57.
population pressure, 42.
PORSILD, 22.
Portage Plains, 15, 16, 18, 21, 66.
predation,
  control by light grazing, 71.
  Franklin ground squirrel, predator
    on duck nests, 71.
  history of predation studies, 114
    (see also crow).
  human disturbance and predation,
    115.
  predator control and waterfowl
    management, 125-127.
predators (see species),
  changes in densities, 120-123.
  movements of and nest predation,
    123-125.
pursuit flights,
  vs. defense flights, 52, 99.
quackgrass, 65-67, 70.
quail, bobwhite,
  sitting on dead eggs, 97.
range reduction,
  and pioneering of ducks, 43.
rate of laying, 95.
redhead,
  ability to pioneer, 42.
  differential age migration, 164-165.
  effect of later hunting season on,
    41.
  habitat, 1.
  hen hiding the brood, 147-148.
  mobility of broods, 144.
  place in bag, 160, 162, 164.
  renesting study by law, 130, 138.
  sex and age ratio in bag, 164.
  sex ratio at hatching, 164.
  sex ratio in spring flight, 164.
  time of laying and nest
    construction, 93.
redstarts, 13.

re-laying,
  in ducks, 135.
  in jackdaw, 133.
renesting, 129-142.
  after loss of brood, 136-137.
  appearance, first nests vs. renests,
    132.
  clutch size, 130-132.
  compensation for early-season nest
    loss, 140.
  continuous laying, 134-136.
  courtship, 98-101, 151.
  distance between nests of
    individuals, 137.
  early studies of, 129-130.
  indeterminate layers, 129.
  and inventory counts, 138-139.
  location of renests, 137.
  of major importance, 141.
  number of unsuccessful hens which
    renest, 137-138.
  persistence, variation between
    species, 139-141.
  re-laying, 135.
  "renesting interval," 132-134.
re-pairing, 99-101, 151.
reproductive potential, 113.
reproductive system of laying hen,
  135, 136.
rest and feeding periods,
  activity during off-nest period,
    102-103.
resting places,
  on arrival, 20-21.
restocking of marshes (see artificial
  stocking).
retrievers (see dogs).
retrieving eggs, 101-102.
RIDDLE, 136.
ring-necked duck,
  place in Delta bag, 160.
river ducks (see species).
robins,
  homing of juveniles, 35.
  migration of, 12.
ROWAN, 168.
ruddy duck,
  habitat, 1.
  increase in kill at Delta, 160.
  mobility of broods, 144.
  place in Delta bag, 160.
Saskatchewan, 114, 123, 165.
scaup, greater,
  place in Delta bag, 160.
scaup, lesser,
  abortive migration attempts, 19.
  eggshell carrying, 107.
  habitat, 1.
  mass movement of, over 20 mile
    front, 167.

place in bag, 162-163.
plumage change in breeders vs.
    non-breeders, 152.
reduction in numbers after World
    War II, 162-163.
sex and age ratio in bag, 164.
sex ratio in spring population, 164.
spring migration, 11.
scientific names, 174-176.
sedges, 66.
SETON, 47.
sex ratio,
    at time of hatching, spring flight,
        and bag, 164.
    composition of fall gatherings of
        pintails, 153.
    of autumn kill, 164.
    of spring flight, 23, 164.
sexual behavior (see behavior).
shooting pressure (see gunning).
shooting season, 151-170.
shoveller (see specific reference).
skunk,
    cause of nest loss, 113-114.
    fluctuation in numbers, 121, 126.
    habitat destruction by light
        grazing, 71.
    human disturbance and skunk
        predation, 115.
    inhabitant of nesting meadows, 71.
    movements of and nest predation,
        122.
    skunk control, 126.
skunk grass, 88.
SMITH and HAWKINS, 138.
snow bunting, 11.
snowberry, 66.
snowy owl, 11.
South Dakota, 43, 70, 138.
sow thistle,
    a dominant plant, 3.
    as nesting cover, 66.
sparrow, song,
    renesting interval in, 133.
species association,
    during migration, 23.
SPRAKE, 97.
spring arrival, 11-24.
    arrival point, 13-16.
    dates, 12.
    main flight, 13.
    mallard, 12-13.
    pair status on arrival, 21-22.
    pintail, 12-13.
    resting and feeding places on,
        20-21.
spring passage (see migration).
squirrel, Franklin ground,
    control of, 126.

fluctuation in numbers, 120-121.
foods of, 120.
habitat, 71.
human disturbance and squirrel
    predation, 115.
movements of and nest predation,
    123.
predator on duck nests, 71, 105,
    113-114.
predator on ducklings, 113.
squirrel, Richardson ground,
    habitat, 71.
    influence on nesting cover, 71.
    relationship to grazing, 71.
    use of burrows by pintails, 116.
STIEVE, 133.
STRESEMANN, 154.
STODDARD, 97.
STOUDT and DAVIS, 138.
stubble feeding (see feeding).
stubble fields,
    as nesting cover, 66, 67.
    preference rating for duck nests,
        67.
stubble flights,
    fall, 155-159.
    history of fall, 155.
    mallard-pintail ratio of, 156.
    predictable flight lanes, 157.
    size of flocks, 158.
    spring, 20-21.
    time of occurrence in fall, 156.
study area,
    description of, 2-3.
survival rate,
    and homing rate, 33.
    in blue-winged teal, 31.
    in gadwall, 31.
    in mallard, 32.
    in pintails, 28-30.
    in shoveller, 30.
swans, whistling,
    migration of, 167.
SWANSON, 97.
teal, blue-winged (see specific
    reference).
teal, green-winged,
    courtship, 22.
    increase in numbers in July at
        Delta, 154.
    marsh area preference evidenced
        in bag, 161.
    place in bag, 160-161.
    reduction of kill after World
        War II, 161.
    sex and age ratio in bag, 164.
    species association, 23.
teasing behavior, 99, 100.

techniques, study, 3-8.
temperature (*see* weather),
    and migration, 17-18.
territory,
    a constituent of home range, 48.
    and absence of aggressive behavior,
        59-60.
    history of, 50, 52.
    in one shoveller hen, 48, 49, 51.
    sharing of "favored areas" by pairs,
        56-58.
    territoriality, 47-63.
    variations in territorial behavior,
        60.
Texas, 1, 22, 130.
THOMAS, 97.
"thwarted" pairs,
    gathering of, 152-153.
TICEHURST (*see* NICE).
TINBERGEN, 50, 109, 146.
tolling,
    of intruder by hen, 147.
transients, 15.
    activity of, 21.
    courtship, 21-22.
    food for, 21.
    spring arrival, 13-17.
    stubble flights of, 155-158.
trapping of hens, 5-6.
TRAUTMAN, 60.
tufted ducks,
    imprinting of, 146.
Utah, 29, 31, 114, 140.
VALIKANGAS, 35.
VAN DEN AKKER and WILSON,
    29, 31.
vegetation,
    control of undesirable, 76.
warblers,
    and territorial behavior, 53.
WARD, 76.
water areas,
    creation of, 21.
    on nesting grounds, 71-73.
    vegetation of, 71-73.
water levels,
    influence on nesting cover, 68-70.
WATSON and LASHLEY, 25.

weasels, long-tailed, 71.
weather,
    cold weather and egg dropping,
        89.
    effect of temperature on behavior
        of ducks, 86-87.
    effect on beginning and length of
        nesting season, 85.
    influence on stubble feeding, 156.
    influence on trapping, 6.
    and spring migration, 17-19.
    and population changes in fall,
        167-168.
    relationship to physiological status
        of ducks, 93.
    temperature and hatchability of
        eggs, 88, 89.
    temperature effect on evening
        flight behavior, 93.
wheat, 3, 66, 155.
whitetop, 3, 65, 66, 67, 69, 70, 145,
    149.
Whitewater Lake, 138, 152.
white-winged scoter,
    place in Delta bag, 160.
wild mustard, 66-67.
WILLIAMS and MARSHALL, 114,
    140.
willow, 66, 68, 70.
WILSON (*see* VAN DEN AKKER
    and WILSON).
wind,
    effect on migration, 20.
    effect on resting ducks, 21.
Winnipeg, Lake, 167.
wintering grounds,
    and home range, 48.
    and pair formation on, 41.
    return to, 25, 26.
Wisconsin, 35.
WITHERBY et al, 96, 99.
wood chuck,
    abundant rodent in nesting
        meadows, 71.
wood ducks,
    renesting, 130.
    transplanted young and return to
        nest, 35.
WORMOLD, 99.